Sulphonation Technology in the Detergent Industry

by

W. Herman de Groot

Kluwer Academic Publishers

Dordrecht / Boston / London

Library of Congress Cataloging-in-Publication Data

Herman de Groot, W.
Sulphonation technology in the detergent industry / by W. Herman de Groot.
p. cm.
Includes index.
ISBN 0-7923-1202-3 (HB : acid free paper)
1. Detergents, Synthetic. 2. Sulphonation. I. Title.
TP990.G68 1991
668'.14--dc20 91-2592

ISBN 0-7923-1202-3

Published by Kluwer Academic Publishers,
P.O. Box 17, 3300 AA Dordrecht, The Netherlands.

Kluwer Academic Publishers incorporates
the publishing programmes of
D. Reidel, Martinus Nijhoff, Dr W. Junk and MTP Press.

Sold and distributed in the U.S.A. and Canada
by Kluwer Academic Publishers,
101 Philip Drive, Norwell, MA 02061, U.S.A.

In all other countries, sold and distributed
by Kluwer Academic Publishers Group,
P.O. Box 322, 3300 AH Dordrecht, The Netherlands.

Disclaimer.
While every effort has been made to ensure the accuracy of the information given in this publication, neither the Author nor the Publisher can in any way accept responsibility for damages or liabilities of any kind which may be claimed to result from the use of this book.

Printed on acid-free paper

Printed in the Netherlands

DEDICATION

This book is dedicated to friends with whom I have worked or still work in the sulphonation field. They live throughout the world and work for a variety of companies. I name them in alphabetical order:

Jorge Aparicio
Mario Ballestra
Jim Begnel
Stefano Cardone
Charley Carroll
Dimitri Corso
Birger Dahl
John Davidson
Joao Ferreira da Fonseca Neto
Norman Forster
Marco Galateri di Genola
Manfred Gehrmann
Dave Hill
Frans van der Hoeven
Ray Hurd
Simon Kellet
Bernard Lemaire
Giorgio Locatelli
Mike de Lucia
Klaas Maat
Miel Marchand
Hans Mik
Franco Moretti
Gianmaria Moroni
Franco Olivieri
Sid Palmer
Roman Pardon
Alfredo Parodi
Luigi Perani
Roy Regan
Sylvere van Renterghem
Manfred Schwartz
Inarto Setiadi
Bob Sharod
Luciano Spadon
S.K. Subbaroyan
Romeo del Vecchio
Antonio Zalaquett

and others too numerous to mention, my heartfelt thanks.

Wim

PREFACE

This book is about Sulph(on)ation Technology in its technical entirety, aiming at superiority in final product quality, raw material utilisation, sustained plant reliability and safety, minimisation of liquid effluent and gaseous emissions; it is about the total quality of the operation. It will be of value to engineers and chemists who are, or will be, involved in the practical daily operation of sulphonation plants or R&D activities. The book can also be used as a tool for the teacher in preparing final year projects in a chemical engineering curriculum.

The book covers sulphonation of alkylbenzenes, primary alcohols, alcohol ethers, alpha-olefins and fatty acid methyl esters, with a strong emphasis on the sulphur-based SO_3/air sulphonation technology.
The first part deals with raw material specifications, hazards, storage, handling and physical properties. In the following section the process chemistry is discussed, indicating main chemical reactions, undesired parallel and consecutive reactions, exothermal heat effects and all other process chemistry data that are relevant for process selection and equipment design. The section about the actual process equipment from the various plant equipment suppliers (Ballestra, Chemithon, Mazzoni, Meccaniche Moderne and Lion Corp.) takes into account the chemical reaction engineering aspects derived from the sulphonation technology processing chemistry.

Product quality, product storage and handling, product safety and physical properties are the contents of the next section.
The effluent handling and exhaust gas treatment of the SO_3/air sulphonation technology are further discussed in detail.
Plant instrumentation and computer control in various degrees of sophistication are described in the next section. Plant housing, space requirements, lay-out and required plant documentation are briefly discussed.

A further chapter of the book describes the 20% oleum and sulphuric acid based sulphonation technology. This part is less extensive, since the SO_3/air process is becoming predominant in the manufacture of detergent actives, taking over from the older oleum and sulphuric acid technologies.

A final chapter outlines the virtues of the sulphonation technology as a "learning paradise" for chemical engineers. All kinds of problems and disciplines come together in one concise operation: highly exothermic reactions in combination with substantial viscosity increase of the reaction products, undesired parallel and consecutive reactions affecting final product quality, environmental problems, hazardous and corrosive chemicals demanding careful selection of plant equipment, design and construction materials.

TABLE OF CONTENTS

ACKNOWLEDGEMENTS

The author is indebted to Joop Blitz, Mike Briggs, Alan Duff, Marc Elenbaas, Dave Hill, Giorgio Locatelli, Franco Moretti, George New, Dave Roberts and notably to John Davidson, who in one way or another, either through their support, discussions or valuable comments helped to create this book.

The author would also like to acknowledge the constant support of Janjaap Blom of Kluwer Academic Publishers, Science and Technology Division.

Eindhoven, 1991
Wim Herman de Groot

ABOUT THE AUTHOR

Wim Herman de Groot worked for more than thirty years for Unilever, mainly in Research and Engineering. Before retiring in 1990 he was Unilever's world-wide expert on sulphonation matters. Since 1984 he has been a part-time professor in the Chemical Engineering Department of the Eindhoven University of Technology, The Netherlands. Professor Herman de Groot is a course director for an annual "Practical Sulphonation Course" in Amsterdam under the auspices of the "Center for Professional Advancement", New Jersey, U.S.A.

LIST OF FIGURES

LIST OF TABLES

LIST OF PROCESS & INSTRUMENTATION DIAGRAMS (BALLESTRA)

1 APPLICATION OF SULPHONATES AS ANIONIC SURFACTANTS IN HOUSEHOLD PRODUCTS

1.1 Introduction

The main constituent in a detergent product is a surfactant, a material containing in a single molecule, a lipophilic oil soluble "tail" (usually an organic molecule with C_{12}-C_{14} chain e.g. RSO_3^-) and a hydrophilic water soluble "head" (usually an anionic e.g. Na^+).
The cleaning process by anionic surfactants (active detergents) is based upon the following phenomena:

i) thorough wetting of substrate and dirt due to reduction of water/substrate and water/dirt surface tension;
ii) removing of dirt from substrate;
iii) maintaining the dirt in a stable solution or suspension.

A variety of organic molecules are used in practice as a base material for detergents, either based on mineral oil, crude oil resources or from natural products. The most important organic feedstocks are: Linear Alkyl Benzene (LAB), Primary Alcohols (PA), Primary Alcohol Ethers (PAE) and Alpha-Olefins (AO).

1.2 Most important sulph(on)ates

(i) LABS (Linear Alkyl Benzene Sulphonate), low molecular weight (230 - 245).

This category anionic is among the major surfactants used in all ranges of household detergent formulations, mainly in dishwashing liquids in combination with other anionic surfactants, for example with Lauryl Ether Sulphate and promoting high detergency foam stability, degreasing capacity, good tolerance for hard water. Common ranges of Active Detergent (AD) levels in liquid detergent products are as follows:

LABS	10 - 15% (30%)
PAS/LES	3 - 5% (10%)

The figures in brackets are maximum values for concentrated products.

(ii) LABS (Linear Alkyl Benzene Sulphonate), high molecular weight (245 - 260).

These are the major anionic surfactants used in all ranges of household detergents formulation, but especially in heavy duty laundry products, sometimes in combination with nonionics, alcohol sulphates of the tallow range and soaps.

Heavy duty NSD powders	High foam	Low foam
- LABS	20 - 30%	5 - 10%
- TAS (tallow alcohol sulphate)		2 - 5%
- Nonionics		2 - 5%
- "Tallow" soap		2 - 5%

(iii) PAS (Primary Alcohol Sulphates)

The PAS category is normally divided in the C_{10}-C_{18} alcohol sulphates comprising coconut fatty alcohol sulphates (low in C_{16}-C_{18}), the C_{12}-C_{14} alcohol sulphates, the so-called "lauryl" alcohol sulphates and the C_{16}-C_{18} range, the "tallow" alcohol sulphates (TAS).

The broad cut (C_{10}-C_{18}) alcohol sulphates represent a compromise of cost/performance in terms of detergency, solubility and foaming properties. This product can partially or totally substitute other anionics in liquid and powder detergent formulations with the advantages of good biodegradability and low "defatting" action which is important for human tissue and delicate natural or synthetic fibres.

The narrow cut (C_{12}-C_{14}) alcohol sulphates find their main application in a wide range of personal care products like shampoos, bubble bath products, tooth pastes, dish washing liquids, delicate products for laundry wash. The ammonia or amine salts of lauryl alcohol sulphates are used in shampoos and bubble bath products. The sodium salt in toothpastes, dish washing liquids and delicate laundry products. The C_{16}-C_{18} alcohol sulphates ("tallow") are used as sodium salts in the formulation of heavy duty laundry products for hand- and machine washing. Their detergency power is up to approximately 10% higher than LAS in a wide range of detergent formulations. Furthermore, TAS shows controlled foam, notably at high temperatures, still keeping the advantage of softness in the wash of sensitive natural- and synthetic fibres. The level of TAS in detergent laundry products will vary for different kinds of products in combination with other surfactants:

	Handwashing (medium foam)	Machine laundry washing (low foam)
- LABS	5 - 10%	2 - 5%
- TAS	10 - 5%	5 - 2%
- Amides/Ni		2 - 3%
- Non-ionics		2 - 4%

(iv) AES (Alcohol Ether Sulphates)

AES (C_{12}-$C_{14}/_{15}$ 2-3 EO) can be considered as the most efficient anionic surfactant in terms of: superior detergency power, good tolerance for water hardness, mildness on hands and fibres. The application therefore is wide: from household to personal care and cosmetic products. Unfortunately, sulphated alcohol ether sulphates show a limited stability to hydrolysis at high temperatures, and this restricts their use in heavy duty laundry powders, where high temperatures occur in the spraydrying process of powder manufacture.

The very high stability to calcium ions permits formulation of liquid detergents with limited or no addition of water "softeners" even in case of use in hard water.
The optimum compromise of ethylene oxide addition ranges between 2 and 3 moles per mole of fatty alcohol to keep good foam levels (higher with low percentage of EO) and solubility/mildness (higher with high percentage of EO).

The most important application worldwide of LES 2-3 EO is in dish washing liquid detergent, generally combined with LAS and in shampoos/bubble baths:

	Liquid dish wash detergent	Shampoos/ bubble baths
- LES (C_{12}-C_{14}/$_{15}$ 2-3 EO)	5 - 10%	10 - 30%
- LABS (low MW)		15 - 20%
- CEA (coconut ethanol amides)	2 - 3%	2 - 3%
- Hydrotopes (STS-SXS) alcohol solvent		1 - 3%
- Other actives (i.e. amphotherics/nonionics)		5 - 10%
- H_2O, perfume, colour preservatives	balance	balance

(v) AOS (Alpha-Olefins Sulphonates)

The Lion Corporation, Japan, are one of the principal producers and users of alpha-olefin sulphonates. In addition to fabric washing powders, they also market fabric washing liquids, shampoos, toothpaste and foam bath products containing this active. In the USA, Minnetonka have utilised AOS in hand cleaners/liquid soaps. AOS is a potential replacement for alkyl benzene sulphonates in dish wash detergent liquids formulations with performance peaking at C_{16} chain length.

(vi) FAMES (Fatty Acid Methyl Esters Sulphonates)

The application of FAMES is under development in various detergent products, but so far the realisation in products on the market is very restricted. The typical C_{16}-C_{18} FAMES shows good surface activity (about 90% compared to LABS), high detergent, dispersing and emulsifying power in hard water, high lime soap dispersion and moderate foam levels. FAMES has a good stability to pH and temperature hydrolysis can be easily spraydried and/or incorporated in detergent bars.

The use of FAMES in detergents, although under investigation and development since more than 15 years, has not yet expanded to high levels, mainly because of the following reasons:

- controversial forecasts about availability of petrochemical feedstocks with related cost comparison vis à vis natural sources;
- availability of sufficient quantity of sulphonation grade methyl esters;
- the process to produce high quality FAMES is generally more complex than that for Alkyl Benzene Sulphonates;
- application know-how is not yet completely available and low FAMES solubility involves some restrictions in application, notably concerning the use in liquid detergent products and low temperature washing cycles.

1.3 Bibliography

1. Moretti G.F. (1989) "Product application of sulph(on)ates as anionic surfactants". Paper presented at Practical Sulphonation Technology Seminar, 18-2O April 1989, The Hague, The Netherlands. Center for Professional Advancement, P.O. Box H, East Brunswick, New Jersey O8816, USA.

2. Falbe J.(editor) (1986) "Surfactants in consumer products. Theory, technology and applications". Springer Verlag-Berlin. Section 5.1.3, "Formulations of household detergents".

3. Shell Detergents Brochures:
-"The use of DOBANIC acids and DOBANOL derivatives in light duty liquids". Technical Bulletin UD/O11 (1984).
-"DOBANOL, sulphates and ethoxysulphates as shampoo bases". Technical Bulletin DI 3.2.2 (1985).

2 A BRIEF DESCRIPTION OF THE SULPHONATION PROCESSES USED FOR THE MANUFACTURE OF ACTIVE DETERGENTS

2.1 Introduction

About 2,500,000 tons of sulph(on)ated anionic surfactants are produced annually in the U.S., Western Europe and Japan. The total annual world production of sulphonated anionic surfactants is estimated at about 4,000,000 tons.

Sulphonation plants are scattered around the globe in units with production capacities varying between 3,000 and 50,000 tons anionic surfactants annually. Assuming an average production capacity of, say, 5,000 tons per year, there are at least 800 operational sulphonation plants around the world.

These plants can be found on the production sites of typical industrial chemical companies and on the premises of companies that produce industrial and household cleaning products.

A variety of sulphonation reagents can be used for the sulphonation reaction: SO_3/air from sulphur burning and subsequent conversion of the SO_2/air formed; SO_3/air from stabilised liquid SO_3 or SO_3 stripped from 65% oleum with dried process air, 20% oleum and chlorosulphonic acid.
The diluted SO_3 in air reacts in gas-liquid contact reactors with the liquid organic phase. The sulphonation reaction is highly exothermic.

It is also technically feasible to use liquid sulphuric acid or 20% oleum as a sulphonation reagent. The reactions can be carried out in various types of liquid - liquid contact reactors and are also highly exothermic.

There are four reasons why SO_3/air raised from sulphur is becoming the predominant sulphonation agent for the manufacture of detergent actives:

(i) Versatility

All kinds of organic feedstocks, like alkylbenzenes, primary alcohols, alcohol ethers, alpha-olefins and fatty acid methyl esters, can be successfully converted with SO_3/air as the sulphonating agent to high-quality sulphonate/sulphate active detergents. Sulphonation agents like sulphuric acid and, to a lesser extent, 20% oleum are becoming less attractive because only alkylbenzene feedstocks can be converted to a good quality alkylbenzene sulphonic acid.
For the other feedstocks, sulphonation with 20% oleum (or sulphuric acid) results in products of unacceptable quality and/or low degrees of conversion.

(ii) Safety

Liquid SO_3, 65% oleum, 20% oleum and concentrated sulphuric acid are hazardous chemicals in transport, handling and storage. The EEC directive on major accident hazards is proposing that if more than 25 tons of SO_3 or its equivalent in oleum are stored on site, all regulations outlined in Annexes of Council directive 82/501/EEC have to be obeyed. Among other requirements, the following points are mandatory:

- it must be demonstrated that the sulphonation is performed safely - including providing details of operator training etc.;
- companies need to provide "notification" to the "member state" competent authorities (in the UK, for example, the "Health and Safety Executive");
- note particularly that those to be informed include "any person outside the establishment liable to be affected by a major accident", who should be appropriately informed of the safety measures to be taken and of the correct behaviour to be adopted in the event of an accident. In other words, if more than 25 tons of SO_3 are stored on site, the neighbours of the factory have to be informed via the local authorities.

All these measures will have an effect on the usage of notably liquid SO_3 and 65% oleum in sulphonation applications.
Sulphur, either in liquid or solid form, is a least hazardous option as a starting material for the production of SO_3.

(iii) Costs

Sulphur is considerably cheaper as a starting material from which to raise SO_3 than liquid SO_3, 65% oleum and 20% oleum. It is also more economical than the three liquid options in transport, handling and storage. Most important of all is the problem of spent acids, resulting from sulphonation with 20% oleum (a dark, 80% strength sulphuric acid, difficult to recycle in the chemical industry) or 65% oleum (a dark, concentrated sulphuric acid, in most countries returnable to producers only at very low prices, sometimes not even compensating for transport costs).
Stabilised liquid SO_3 requires very precise temperature control and its residues, which contain noxious stabiliser residues after evaporation, are difficult to handle and dispose of.

(iv) Availability

Liquid SO_3, 65% and 20% oleum and even sulphuric acid are not manufactured in all regions of the globe. In many parts of Africa and Asia sulphuric acid and different derived oleum qualities are not available. Even when sulphuric acid is produced, oleum of various strengths may not be manufactured.

The 20-30% strength oleum is made in an absorption tower where SO_3/air (about 10% SO_3 in air raised from sulphur) is washed with concentrated sulphuric acid. The 40% strength is made in the same way, except that two absorption towers are required.

The absorption rate drops as the concentration of SO_3 in oleum increases. The pressure of SO_3 above fuming sulphuric acid in mm Hg at 35°C is:

20% oleum	1.7	mm Hg
30% oleum	8.2	mm Hg
40% oleum	64.0	mm Hg
65% oleum	214.9	mm Hg

If SO_3/air with 10 volume % SO_3 is raised, the SO_3 partial pressure would be 0.1 x 760 = 76 mm Hg. The driving force for absorption will become very small when the concentration of 40% oleum is approached. 65% oleum cannot be made by absorption of SO_3 from SO_3/air mixtures obtained from sulphur burning, because the required partial pressure of SO_3 in air has to be higher than 214.9 mm Hg. The process to manufacture 65% oleum is therefore in practice as follows:
SO_3 is boiled off from 30% oleum and condensed to liquid SO_3. The liquid SO_3 is subsequently mixed with 20% oleum. This is a completely different technology compared with the absorption process for 20 - 30% oleum manufacture.

2.2 Sulphur-based SO_3/air sulphonation

Figure 1 illustrates the overall block diagram. To raise SO_3/air with a volumetric content between 4 and 7% SO_3, the process air should be dried to prevent the formation of sulphuric acid mist. Filtered ambient air is compressed (about 0.6 bar gauge), chilled (about +5°C) to remove the major part of water by condensation and subsequently dried with a desiccant (silica gel, Alumina) to arrive at a process air dewpoint of about -60°C.

Molten sulphur (150°C) from storage is pumped to the sulphur furnace where sulphur is converted with an excess of oxygen from the process air to SO_2 (4-7% SO_2 by volume in "air"). The furnace outlet temperature of the SO_2/air varies with the percentage of SO_2 in air between 600 and 700°C, indicating the strong exothermic character of the reaction. The SO_2/air flow is cooled in an indirect air cooler from 600-700°C to about 420°C.

SO_2 is converted to SO_3 in the so-called converter tower filled with 4 packed beds of V_2O_5 catalyst on a silica carrier. The reaction is highly exothermic and intermediate cooling of the process gas flow between the various beds with indirect air coolers is required.

Notwithstanding the low process air dewpoint, some sulphuric acid/oleum mist condenses in the coolers following the converter tower at temperatures of about 45-50°C. This highly reactive mist can affect the quality of the subsequent sulphonation reaction and therefore a high-efficiency demister is installed before the actual sulphonation step.

The sulphonation reaction can be characterised as follows:

- the reaction between SO_3 and organic feedstock is almost instantaneous;
- undesired parallel and consecutive reactions may occur, notably at higher temperatures;
- the reaction is highly exothermic;
- the viscosity of the organic acid (500-1000 cP) is significantly higher than the viscosity of the organic feedstock (5-10 cP).

Figure 1 Sulphonation-sulphation plant with sulphur burning

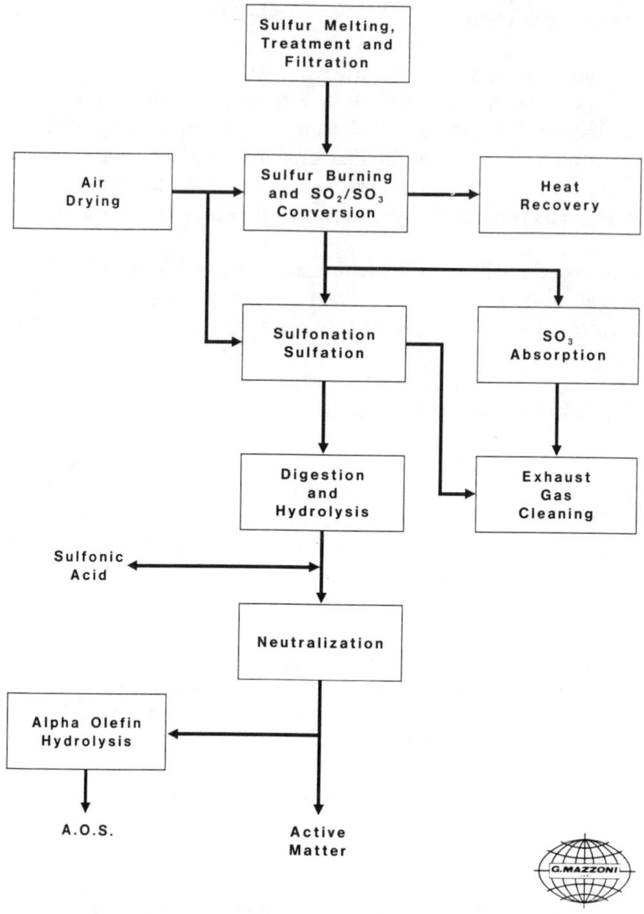

Falling-film reactors of different design are nowadays widely used for the sulphonation reaction.

After the reactor, the SO_3 exhausted gas is separated from the organic acid. The exhaust gas, containing small amounts of non-converted SO_2, unreacted SO_3 and some entrained organic acid, has to be cleaned before emission to ambient atmosphere. The organic aerosol and fine SO_3/H_2SO_4 droplets are separated from the exhaust gas flow in an electrostatic precipitator (ESP) and the gaseous SO_2 and traces of SO_3 gas are washed from the process air in a scrubber by dilute caustic solution, thus producing a mixed sulphite/sulphate solution.

Depending on type of organic feedstock and consequent organic acid, further reaction steps may be required before organic acid neutralisation. LAB sulphonic acid has to pass an ageing step for conversion of an intermediate product to the desired LABSA. Moreover, a hydrolysis or stabilisation step is required to convert the formed anhydride to LABSA with a small amount of water (appr. 1% on LABSA). Alcohol and alcohol ethoxylate sulphonic acids, on the other hand, must be neutralised immediately after sulphonation to avoid undesired by-products, formed in consecutive reactions. After ageing and hydrolysis, LABSA is a stable product, which can be stored and transported.

The neutralisation reactions can be carried out with many alkaline chemicals like caustic, ammonia and sodium carbonate. The reaction with diluted caustic to a paste containing between 40 and 70% AD, depending on organic acid type, is most widely used. Neutralisation with diluted caustic can be characterised as follows:

- the reaction between organic acid and caustic is instantaneous;
- undesired reactions may occur if micro-dispersion of organic acid in the diluted caustic phase fails;
- gel may be formed at high temperatures;
- the reaction is highly exothermic;
- the AD paste shows non-newtonian rheology characteristics, so-called pseudoplastic behaviour. The viscosity drops at high shear rates.

Various loop-type reactors, consisting of a circulation pump, homogeniser (where the acid is introduced in the circulating alkaline paste) and heat exchanger, are used for the complex neutralisation step.

From this brief description of the total SO_3/air sulphonation operation the following points come to light:

- the SO_3/air gas-raising plant is an example of a "heavy" chemical industry operation. Highly corrosive and hazardous chemicals like SO_2, SO_3 and SO_3/oleum/sulphuric acid mist are produced at elevated temperatures. This kind of process is preferably carried out in a fully continuous mode, 24 hours per day, 7 days per week;
- the sulphonation and neutralisation reactions themselves are delicate in the sense that inaccurate operations lead to undesired by-products, bad colours and poor yields of converted organic feedstocks;
- poorly controlled operations may cause hazardous situations to people on site and in the adjacent surroundings;
- mistakes and poor operational standards will not only result in poor product quality and poor yields, but will also reduce seriously the life-span of the total plant.

2.3 Sulphonation with 20% oleum

Continuous sulphonation with oleum is a four-stage process (see figure 2).

First, alkylbenzene is reacted with a 10 to 15% excess (by weight) of 20% oleum. The reaction is carried out in a loop-type reactor consisting of a centrifugal circulation pump of high capacity, where fresh alkylbenzene and 20% oleum are mixed in the circulating reaction mixture, a heat exchanger to maintain the temperature at about 50 °C and a product off-take point. The resulting product - a mixture of sulphonated material and residual sulphuric acid - then passes to the "hold-up coil", where any unsulphonated alkylbenzene has time to react. In a second loop reactor the sulphonated product is then diluted with water to reduce the strength of the sulphuric acid to about 75 - 80%. Much of the heat evolved during the dilution is removed in the dilution loop or diluter.

Next, the dilution product is allowed to separate into two phases and the lower, aqueous acid layer is removed as "spent acid".

Figure 2 20% oleum sulphonation, a four-stage process

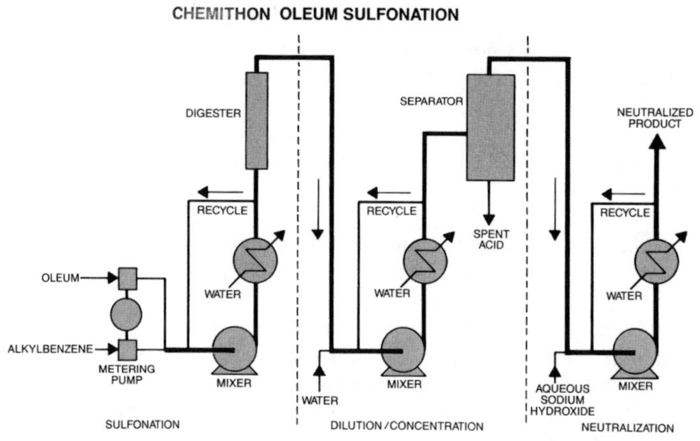

Much of the colour produced during sulphonation is contained in the spent acid. The upper layer in the separator, which contains around 90% alkylbenzene sulphonic acid (the remainder being primarily dissolved sulphuric acid) is also removed from the separator and neutralised with sodium hydroxide solution or with sodium carbonate solution to yield the active paste. When sodium hydroxide is used a considerable amount of heat must be removed. This is done in the third loop reactor of this process, the neutralisation heat exchanger (also known as the neutraliser). If sodium carbonate solution is used, the neutralisation is much less exothermic and no heat exchanger is needed at this stage.

The active paste has a typical composition of 40 - 50% AD, 5 - 9% sodium sulphate and less than 1% unsulphonated organic material (free oil). Pastes from oleum sulphonation do not require bleaching.

2.4 Sulphonation with sulphuric acid

The chemical reaction is as follows:

$$RH + H_2SO_4 \rightarrow R\,SO_3H + H_2O$$

As can be seen, water is set free by the reaction. The presence of water retards the sulphonation reaction, or even prevents it altogether. This is the reason why a very large excess of H_2SO_4 is used to bring the reaction to completion. (The oleum excess is used for the same reason.)

Ideal reaction conditions for a batch process are:

H_2SO_4/AB ratio	1.6 - 1.8
total reaction time	1.0 - 1.5 h.
temperature	55°C

Dilution, separation and neutralisation can take place in the same reactor or several batch units may be used for the consecutive steps (see also 2.3). Sulphuric acid sulphonation in a continuous loop reactor system is feasible when an H_2SO_4/AB ratio of at least 1.80 is applied. In this case, as well as when 20% oleum is used, reasonably short reaction times are sufficient to complete the reaction. With increasing H_2SO_4/AB ratio, the amount of dark 80% sulphuric acid (spent acid) will increase proportionally.
Industrial sulphuric acid sulphonation of LAB for the production of active detergents is hardly applied any more.

2.5 Sulph(on)ation with chlorosulphonic acid

Chlorosulphonic acid, particularly with batch operation, is best suited for production of a range of products on a relatively small scale. Chlorosulphonic acid is still used for the sulphonation of fatty alcohols, fatty alcohol-ethoxylates and related detergent raw materials with OH-groups available for the attachment of a SO_3H group.
The reaction of Lauryl alcohol with chlorosulphonic acid illustrates for example the chemistry involved:

$$C_{12}H_{25}OH \quad + \quad ClSO_3H \quad \rightarrow \quad C_{12}H_{25}OSO_3H \quad + \quad HCL$$

The hydrogen chloride must be absorbed to give a 30% HCl solution as a by-product.

Sulphonation with chlorosulphonic acid requires special corrosion-proof equipment either glass-lined steel or all glass. The latter is especially suitable for small batch size sulphonation. The HCl absorber, too, is built either of glass-lined steel or is all glass.

Bozzetto (Bergamo, Italy) offers a continuous chlorosulphonic acid sulphonation process which comprises two water-jacketed glass vessels for sulphonation and neutralisation, and a HCl absorption column. Organic feedstock and chlorosulphonic acid are mixed on a rotating disc. Under the centrifugal action of the disc, the reaction mixture is sprayed as a thin film on to the wall of the reaction vessel. The acid product falls to the base of the reactor and then onto a similar rotating disc system, where it is mixed with alkali and sprayed onto the wall of the neutralisation vessel. The unit is operated under slightly reduced pressure to remove HCl gas from the reaction zone. The gas is absorbed in the column to produce 35% hydrochloric acid.

Industrial sulphation with chlorosulphonic acid is only applied for the manufacture of relatively small quantities of detergent specialities.

2.6 Bibliography

1. Falbe J. (editor) (1986) "Surfactants in consumer products. Theory, technology and applications". Springer Verlag, Berlin. Section 3.2.1.1. "General production technologies". pp 54-62.

2. Davidsohn A., Milwidsky B.M. (1978) Synthetic Detergents. (6th edition) John Wiley, New York.
pp 122-137 "Sulphonation with sulphuric acid and/or oleum", "sulphonation with SO_3", "sulph(on)ation with chlorosulphonic acid".

3. CHEMITHON BROCHURE (1988).
5430 West Marginal Way S.W., Seattle WA 98106-USA.

4. European Commission-Implementation of the directive on major accident hazards of certain industrial activities. Article 19 review of annexes 1,2,3-proposed by UK on SULPHUR TRIOXIDE (1985), Health and Safety Executive, 25 Chapel Street, London NW1.

3 RAW MATERIALS QUALITY, STORAGE, HANDLING AND SAFETY

This section deals with all major aspects which require consideration when proposing to use the common raw materials found in sulphonation.

3.1 Sulphur

Sulphur crystallises in at least two distinct forms, rhombic and monoclinic. Rhombic sulphur, SA, is stable at atmospheric pressure up to 95.5°C at which temperature transition to monoclinic sulphur, SB, can take place. Monoclinic sulphur is stable up to its natural melting point of 114.5°C.

The molecular constitution of liquid sulphur undergoes significant and reversible changes with increasing temperature, as evidenced by the characteristic temperature dependance of physical properties, notably viscosity. The sudden increase in sulphur viscosity above 159°C is attributed to the formation of polymeric sulphur molecules. (see 3.1.4 and figure 3.)

3.1.1 Sulphur quality

Sulphur can be purchased to a wide variety of specifications and standards. The quality, as defined by the level of impurities, depends upon the source and method of production. Commonly sulphur is obtained from the extraction of native deposits found in unconsolidated volcanic rocks, recovered from hydrogen sulphide present in sour natural gases or produced by the hydrodesulphurisation of crude oil fractions in refinery operations.

The level of impurities in the sulphur used in sulphonation operations should normally not exceed 0.5% and a target minimum sulphur content of 99.7% is preferred. The types of impurity and their effect on sulphonation operations are summarised below. Generally, sulphur analysis is performed very infrequently but in cases where new suppliers appear or processing problems occur, it is advisable to have the sulphur quality checked to ensure that the specification is being met. Because of the low specified levels of some of the impurities, the expertise available in specialised laboratories should be employed for this analysis.

Impurity	Maximum level in ppm (Test Method)	Origin & consequences
Ash	500 (BS4113)	Contaminants picked up in transit. Sludge contamination of heated surfaces requiring manual removal. Physical coating or choking of the catalyst bed or pre-bed filter, increasing pressure drop requiring manual cleaning.
Acids (as H_2SO_4)	100 (BS4113)	Acids are formed slowly when sulphur is exposed to moisture or air. Corrosion and reduced equipment life will result, in addition to premature catalyst deactivation.
Water	100 (BS4113)	See acids.

Hydrocarbon	200 (BS4113)	Hydrocarbons increase the amount of water vapour in the sulphur burner, giving acids. Organic sulphides which can spontaneously ignite may be formed in the presence of sulphuric acid. Hydrocarbons in the form of bitumenous compounds necessitate more frequent cleaning of heating surfaces and can block pump ball valves.
Iron	5.0 (BS4113)	Iron can have adverse effects on the sulphur filtering process.
Arsenic	0.25 (BS4113)	Arsenic will poison the vanadium catalyst reducing its service life and requiring premature replacement.
Selenium	2.0 (BS4113)	Similar to arsenic.
Tellurium	2.0	Similar to arsenic.
Fluorine/ chlorine	Not detectable	Fluorine and chlorine cause catalyst damage by reaction with silica.
Hydrogen sulphide	Not detectable	Hydrogen sulphide is not only toxic but, if allowed to accumulate in storage systems, can give rise to explosion hazards.

3.1.2 Sulphur storage, handling and safety

This section should be read in conjunction with the raw material hazard data sheet (see 3.1.3.) and the physical property data sheet (see 3.1.4.).

Solid sulphur

Solid elemental sulphur is virtually non-toxic although sulphur dust can cause systemic poisoning if inhaled and can irritate the eyes and mucous membranes of the respiratory tract. No specific exposure limits exist for solid sulphur and therefore it should be considered as a nuisance dust with the following exposure limits:

- 10 mg/m^3 (total dust);
- 5 mg/m^3 respirable dust.

Adequate ventilation is required.

Because of its low ignition energy and tendency to develop static charges sulphur presents a fire and explosion hazard. In the event of fire, sulphur burns with a slow blue flame, invisible to the naked eye in daylight, and evolves sulphur dioxide (see 3.2).

During melting of sulphur, hydrogen sulphide originating from hydrocarbon impurity in the raw material may be evolved. Hydrogen sulphide not only has an extremely high acute toxicity and is an irritant to the eyes and respiratory tract but also has a relatively low explosive limit (4.3 vol. % at 20°C, 3.4 vol. % at 132°C). Simple tests (Draeger tube) are available to check for the presence of hydrogen sulphide in areas where workers may be exposed. Although hydrogen sulphide is often initially detected by its characteristic "rotten eggs" smell, such indications are unreliable as it has a deceptively sweet smell at concentrations between 40 and 130 mg/m^3 and deadens the sense of smell above this range. For a material as toxic as hydrogen sulphide local legislation defining maximum allowable exposure limits should be available. In the absence of local regulations the UK (HSE) or USA (ACIGH) standards should be applied.
These are:

- 14 mg/m^3 (10 ppm); 8 hour time-weighted average;
- 21 mg/m^3 (15 ppm); 10 minute time-weighted average.

In areas where exposure to hydrogen sulphide is possible, e.g. sulphur melting pits, adequate ventilation is required.

Solid sulphur should be stored in a cool, clean, dry environment; protection from exposure to rain or snow is essential to avoid the formation of corrosive acids. In the construction all ledges and other horizontal surfaces should be avoided. Sulphur should not be stored with chlorates, nitrates or other oxidising materials, mineral acids or metal powders. All sources of ignition must be excluded from the immediate vicinity of storage areas and a "permit to work" system is necessary if welding is required. For enclosed storage, effective earthing is required. Mild steel construction is preferred and copper, silver or their alloys should not be used in direct contact with sulphur as it readily reacts with these materials at ambient temperatures.

Solid sulphur should be handled in such a way as to prevent the formation of dust clouds. As sulphur dust in air ignites easily, explosions may occur (explosive limit is 3.5 g/m^3) potentially resulting in ignition of the material stockpile. Damage may be serious if the explosion occurs in a confined area. If solid sulphur is to be handled in mechanical systems e.g. elevators, conveyors, they should be of the type which avoids the possibility of a build-up of frictional heat, they should be enclosed and be provided with explosion relief vents. Conveyors and bucket elevators should generally be constructed of non-ferrous materials and metallic parts electrically bonded together and earthed to avoid static discharge. Electrical drive gear must be flameproof (Class II Zone 1) and electrical light fittings should be dust-tight.

Liquid Sulphur

The primary hazards of liquid sulphur are its relatively low ignition point, the possible presence of hydrogen sulphide (see solid sulphur) and the fact that it is a hot liquid capable of producing severe burns.

The flash-point of liquid sulphur varies with purity and the test method, but is always higher than the normal handling temperatures (121-154°C). Pure sulphur has a flash-point of 188°C while for relatively impure crude sulphur containing higher than specified levels of hydrocarbons the flash-point may reduce to 168°C. The vapour concentration above pure sulphur, free of hydrocarbons or hydrogen sulphide in the normal temperature range is too low to form an explosive mixture with air.

Liquid sulphur can cause severe burns. First aid treatment involves immediate cooling of the affected area after which the sulphur crust should not be removed as it acts as a sterile dressing. Further treatment should be administered by a physician.

Molten sulphur is transported and stored in mild steel vessels. Agitation, which can release trapped hydrogen sulphide, should be avoided during loading and unloading. Consequently liquid sulphur storage tanks should be designed with filling lines which extend near to the tank bottom. Storage tanks should be vented to atmosphere preferably using at least two steam-jacketed vent lines. Heating coils with steam at a temperature close to 150°C need to be provided inside the storage tanks, an external heating coil on the roof and all pipework, pumps and valves must be jacketed and insulated. Piping kept full of molten sulphur will last longer than piping which is drained and allowed to cool when not in service in which sulphuric acid may form, resulting in corrosion. Tanks and pipework should be electrically earthed. It is necessary to equip storage tanks with steam blanketing to be used in case of fire and preferably to incorporate a high temperature alarm in the upper section of the tank to warn of abnormal situations.

3.1.3 Sulphur Hazard Data

Raw material hazard data sheet (see next page)

During transfer of liquid sulphur, particularly into the storage tank, personnel should be completely protected by clothing including safety boots, heat resistant gloves and safety glasses. Full-face visors should be worn when disconnecting pipework used for transporting liquid sulphur. Several systems are used for discharging tankers or rail cars containing liquid sulphur. In general the tanker must be electrically grounded and it is preferable to minimise transfer distances through the steam jacketed pipework, discharge arrangements are often constrained by the source and nature of supply. When transferring sulphur to the process it is considered advantageous if the flow from the storage tank to the intermediate pump supply tank is by gravity. It is however necessary to filter the sulphur prior to use in the process to remove iron and bitumenous contaminants.

Centrifugal, gear type and reciprocating piston pumps can be used for handling molten sulphur. Centrifugal pumps should be vertical shaft sump pumps without stuffing boxes and the pump housing is generally submerged in the molten sulphur. Gear pumps are used to transfer or meter molten sulphur. Generally, ordinary iron or steel pumps wear rapidly, although wear can be minimised by facing critical parts with hard, abrasion-resistant materials. Piston pumps do not wear as rapidly as gear pumps but the presence of hydrocarbon impurity may cause the piston to jam or sticking of check valves. Filters to protect the pump must always be used.

RAW MATERIAL HAZARD DATA SHEET

CHEMICAL NAME	PHYSICAL FORM
Sulphur (S)	Crystals, powder or liquid

FIRE/EXPLOSION HAZARDS
Combustible, dust may produce risk of explosion with air.

REACTIVITY
When heated gives toxic gases (H_2S).

HEALTH
Can cause irritation to eyes, skin and if inhaled. Molten sulphur causes burns.

PROTECTION/HANDLING REQUIREMENTS

FACE/EYES	HANDS	BODY	FEET
Wear chemical workers goggles	Wear gloves	Wear overall	Wear normal factory safety footwear

	EMERGENCY ACTIONS			
	INHALATION	EYES	SKIN	INGESTION
FIRST AID	Remove from source of dust	Wash well with water	Wash skin with water	Drink water or or milk Med. att.
FIRE	Fight fires with water spray or stream nozzles. Formation of toxic SO_2 gas. Protective clothing, see above.			
	CONTAIN/ABSORB/ADSORB		FLUSH/WASH AWAY	
SPILLAGE LEAK	Sweep up and collect spillage into suitable sealable container. Protective clothing, see above.		Not applicable.	

3.1.4 ENGINEERING DATA - PHYSICAL PROPERTY DATA SHEET

NAME: SULPHUR		MOLECULAR WEIGHT: 32.064				
CHEMICAL FORMULA:		S (Rhombic - SA; Monoclinic SB)				
APPEARANCE: Solid:		Fine yellow flakes; coarse/fine mixture of broken lumps				
Liquid:		Dark brown and viscous				
DENSITY: Liquid	kg/m^3	1799	1795	1787	1778	1770
TEMPERATURE:	°C	125	130	140	150	159
VISCOSITY	Ns/m^2	0.011	0.0067	0.03	0.5	93
(see also fig. 2)						
TEMPERATURE:	°C	120	157	160	165	187
SOLUBILITY (WATER):		g/l	INSOLUBLE			
MELTING POINT:	°C	SA	110.2			
		SB	114.5			
BOILING POINT:	°C		444.6			
VAPOUR PRESSURE:						
(liquid)	mbar	1.33	6.65	13.3	133.4	1013.1
TEMPERATURE:	°C	183.8	223.0	243.8	327.2	444.6
FLASH POINT:	°C	188 (pure) 168 (impure)				
SPECIFIC HEAT CAPACITY (liquids):	kJ/kg°C	(0.706 + 0.65 (T+273)/1000 e.g. at 119°C Cp = 0.96				
THERMAL CONDUCTIVITY:	W/m°C	1.54×10^{-5} (at 100°C)				
HEAT OF VAPOURISATION:	kJ/kg	290.1 (at 440°C including of dissociation of S_2)				
HEAT OF FUSION:	kJ/kg	SA to liquid 49.8 SB to liquid 38.5				
HEAT OF COMBUSTION (25°C)	kJ/kg of S	-9230				
IGNITION TEMPERATURE LIQUID SULPHUR	°C	appr. 250				
DUST CLOUD AUTO IGNITION TEMPERATURE	°C	appr. 190				

Figure 3 Viscosity of sulphur

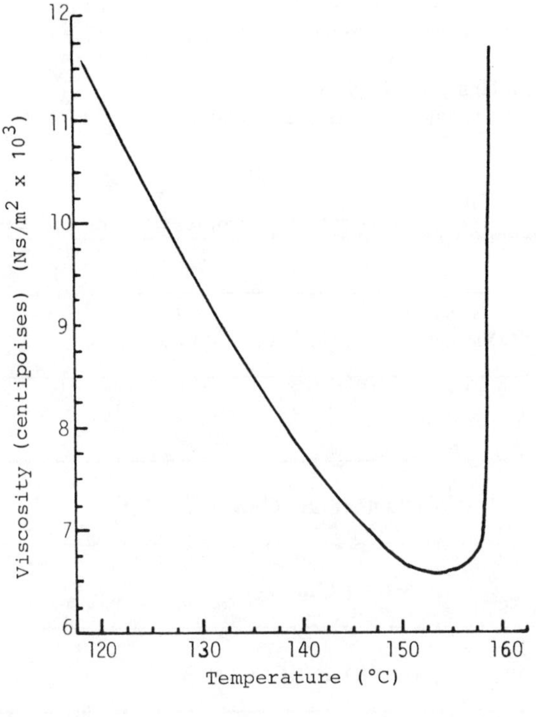

3.2 Sulphur dioxide

Sulphur dioxide gas is produced in the process from the combustion of sulphur under controlled conditions. Exposure to sulphur dioxide gas may occur due to leaks in the process or if the sulphur store accidentally catches fire. Leaks can be detected by placing the stopper of a bottle containing ammonium hydroxide close to the area of the suspected leak. Dense white fumes are released by the reaction of sulphur dioxide with ammonia. Continuous cycle sequential monitoring systems e.g. Gelman type are recommended for the early detection of sulphur dioxide leaks, particularly if the plant is located inside a building.

3.2.1 Hazard data

In the event of a sulphur fire, large quantities of sulphur dioxide may be released. The procedures for extinguishing the fire outlined in section 3.1.2 should be followed. It is important to stay up-wind of any sulphur dioxide source, away from low lying areas where the gas could collect. Any closed space should be well ventilated before entering and self- contained, positive pressure breathing apparatus worn. In the event of exposure to sulphur dioxide breathing may become difficult. Oxygen must however only be administered by a qualified physician.

Sulphur dioxide is extremely irritating. A concentration of 0.3 to 1 ppm can be detected by the average individual, 3 ppm has an easily noticeable odour and 20 ppm is the least amount which is irritating to the eyes. In the absence of local legislation the UK (HSE) or USA (ACIGH) covering maximum allowable exposure limits should be applied:

- 5 mg/m^3 (2 ppm); 8 hour time-weighted average;
- 13 mg/m^3 (5 ppm); 10 minute time-weighted average.

3.2.2 ENGINEERING DATA - PHYSICAL PROPERTY DATA SHEET

NAME: SULPHUR DIOXIDE	MOLECULAR WEIGHT: 64				
CHEMICAL FORMULA:	SO_2				
APPEARANCE:	COLOURLESS GAS				
SPECIFIC GRAVITY:	2.264 w.r.t. air				
SOLUBILITY (WATER):	g/l	185 at 0°C: 43 at 50°C			
MELTING POINT:	°C	-75.5			
BOILING POINT:	°C	-10.0			
SPECIFIC HEAT CAPACITY	kJ/kg°C	0.645			
HEAT OF REACTION TO SO_2: $S + O_2 \rightarrow SO_2 + 9230$ kJ/kg s	kJ/kg s	-9230 (base 25°C)			
VISCOSITY:	Ns/m^2	20×10^{-6}	28×10^{-6}		
AT TEMPERATURE:	°C	200	400		
THERMAL CONDUCTIVITY:	W/m°C	1.16×10^{-2}	1.45×10^{-2}	1.74×10^{-2}	2.33×10^{-2}
AT TEMPERATURE:	°C	225	350	450	650

3.3 Sulphur trioxide

Sulphur trioxide is produced by the catalytic oxidation of sulphur dioxide. Exposure to sulphur trioxide may occur due to leaks in pipes and process equipment. On escape to the atmosphere, sulphur trioxide reacts with water to form sub-micron particles of sulphuric acid "smoke" which may be sufficiently dense to obscure the source of the leak.

There exist three solid forms of SO_3 called A, B and C. Only the A-form is stable, while B and C are not stable. The melting point of the A-form is 62°C, at which temperature the vapour pressure of liquid SO_3 is 2.5 atmospheres. However, it appears that the A-form never crystallises from the liquid phase, that only forms if B crystals are present initially. These B-form crystals will in time (which might be very long) transform into A, therefore it is essential to prevent the formation of B crystals.

Liquid SO_3 is relatively stable between 32°C and 44.5°C at 1 atm. Below 32°C the B-form starts crystallising and the danger then exists that B crystals might transform into the A-form, which cannot be melted without increasing the pressure to 2.5 atm, as noted above.

The stabilisers which are used in liquid SO_3 only serve to prevent the crystallisation of the B-form from liquid SO_3 below 32°C. If the temperature drops further to 16.5°C, the C-form will crystallise, though there does not appear to be much danger of the C-form transforming directly into A. Stabilisers in liquid SO_3 are patented inhibitors, such as boron compounds, methane sulfonyl chloride and phosphorous oxychloride.

Sulphur trioxide is normally kept in the liquid state at a temperature of about 35°C and ambient pressure. The vapour pressure at 35°C is about 410 mm Hg. The commercial sulphan has a melting point of 16.8°C and boiling point of 45°C.

3.3.1 Hazards of sulphur trioxide

Sulphur trioxide reacts with water to form sulphuric acid. Exposure to SO_3 is therefore exposure to sulphuric acid mist which intensely irritates the mucous membranes, including the eyes and respiratory tract lining. Exposure to high concentrations of sulphuric acid mist (in excess of 100 mg/m^3) can lead to death. Sensitivity to sulphuric acid aerosol varies with individuals. Normally 0.125 ppm to 0.5 ppm may be mildly annoying, 1.5 - 2.5 ppm definitely unpleasant and 10 -20 ppm unbearable. No time-weighted average exposure limit has been established for SO_3, but the standard adopted by most regulatory authorities for exposure to sulphuric acid aerosol is 1 mg/m^3, 8 hours time-weighted average.

The failure of a filling or unloading line attached to a road tanker of, say, 18 ton, capacity has been identified as an event which is representative of the foreseeable sizes of release from a typical installation. Severance of the line, e.g. by tanker moving while it is being filled or unloaded, would cause the release of sulphur trioxide. The SO_3 released will spread in a pool, reacting violently with any water present, either in or on the ground, until all the water is used up. The result will be the production of a large cloud of dense sulphuric acid mist. Once the readily available water has been used up, the pool of sulphur trioxide will continue to emit sulphuric acid mist fumes from reaction with moisture in the air. This will continue until all the sulphur trioxide is used up or remedial measures have been taken. Trials by ICI with 65% oleum have shown that even with the immediate treatment of a spill, more than half of the spilled material will still be lost to the atmosphere.

It has been calculated that a spill during tanker filling or unloading operation as outlined above could release approx. 1.0 ton of SO_3 as sulphuric acid mist, which would form the initial cloud. As the cloud is blown along by the wind, it will be diluted by entrained air until the concentration is no longer hazardous. The distance to this point will depend on the wind and weather conditions. In stable weather conditions (Pasquill category D and at 5 m/s windspeed) a dispersion model predicts that the clouds could be hazardous up to about 1 kilometre from the point of release!

A cloud will pass a 1 kilometre point in about 4 - 5 minutes under these weather conditions. During this time only a fraction of the concentration in the cloud will penetrate into a building and, because of this, people sheltering inside would be afforded some protection. However, the closer the building is to the spill, the higher the concentration of sulphuric acid mist in the cloud and the higher the dose of sulphuric acid mist which would be received during the passage of the cloud. At some point nearer to the spill the concentration inside the building could rise to such a level that the received dose of sulphuric acid would be very serious.

It has been calculated that this is possible within about 500 metres from the spill in the conditions outlined above.
People unable to seek protection indoors or especially susceptible to irritant gases or those who cannot easily undertake emergency action would be at serious risk from release of this size up to 1 kilometre from the release. In more stable weather, at slower wind speeds or with larger spills, there would be a serious hazard to people at greater distances.
Serious incidents have occurred with the spillage of oleum or SO_3. A serious incident occurred, for example, in August 1973 at Carton, California, U.S.A. It involved about 1400 ton of oleum and sent clouds of acid over a 56 square kilometre area of scarcely populated land. Local hospitals treated 38 persons exposed to the potentially toxic fumes. The incident occurred when a fitting ruptured on a pipe outside an oleum tank. It took 4.5 hours to plug.

In the UK there have been several serious incidents involving the loss of SO_3. One of the incidents occurred in March 1977 at a chemical manufacturers' in the south-east. Less than half a ton of SO_3 was lost in one or two minutes when a flexible 2" diameter pipe connecting a road tanker to a stock tank split. Fortunately the weather conditions were fairly unstable, with a strong wind, and this aided dispersion. The fumes stayed in a narrow plume and travelled at least 1.5 kilometres before reaching houses. The intervening country-side was largely unpopulated, but people at a riding school at 900 metres from the spill were affected and required treatment.

In 1982 there was a smaller-scale incident in the Midlands involving explosion of drums containing liquid SO_3 and oleum 65. It was believed that water entered the drums causing overpressure. The incident caused a great deal of local alarm and national publicity, but damage was confined to vegetation in the nearby fields.

In conclusion, past experience and an assessment of the hazards and risks indicate that liquid SO_3 and various strengths of oleum have the potential to present a major accident hazard. The European Commission has proposed that these chemicals should be included in Annex II (isolated storage) and Annex III (process activity) of the EC Directive. In the case of Annex II for the application of Articles 3 and 4 a threshold quantity of 15 tons is proposed.

<table>
<tr><td colspan="5">RAW MATERIAL HAZARD DATA SHEET</td></tr>
<tr><td colspan="2">CHEMICAL NAME
Sulphur trioxide (SO_3)</td><td colspan="3">PHYSICAL FORM
Gas or liquid</td></tr>
<tr><td colspan="5">FIRE/EXPLOSION HAZARDS
Non-flammable, explosion risk on contact with water.
REACTIVITY
Reacts violently with water (forming H_2SO_4) and alkaline materials, mixed with air forms corrosive mist/fume.
HEALTH
Fume/mist causes severe irritation if inhaled.</td></tr>
<tr><td colspan="5">PROTECTION/HANDLING REQUIREMENTS</td></tr>
<tr><td>FACE/EYES</td><td>HANDS</td><td>BODY</td><td colspan="2">FEET</td></tr>
<tr><td>Wear full face mask</td><td>Long vinyl or neoprene gloves</td><td>Vinyl and neoprene protective work clothing</td><td colspan="2">Wear acid/chem. resistant boots (vinyl or neoprene)</td></tr>
</table>

<table>
<tr><td></td><td colspan="4">EMERGENCY ACTIONS</td></tr>
<tr><td></td><td>INHALATION</td><td>EYES</td><td>SKIN</td><td>INGESTION</td></tr>
<tr><td>FIRST AID</td><td>Remove from source of fumes/mist

Med. att.</td><td>Wash with copious amount of water

Med. att.</td><td>Wash skin with water

Med. att.</td><td>Drink two glasses water or milk

Med. att.</td></tr>
<tr><td>FIRE</td><td colspan="4">Avoid contact with water. Keep storage containers cool with water. When fighting fires, wear suitable protective clothing (see above) and wear supplied air or self-contained breathing apparaturs.</td></tr>
<tr><td rowspan="2">SPILLAGE LEAK</td><td colspan="2">CONTAIN/ABSORB/ADSORB</td><td colspan="2">FLUSH/WASH AWAY</td></tr>
<tr><td colspan="2">Treat with an excess (4 times or more) of anhydrous sodium sulphate. Wear protective clothing and supplied air or self-contained breathing apparatus.</td><td colspan="2">Residual amounts with large volumes of water - using a Fog nozzle. Do not direct Jet of water to spillage, inform authorities. Wear protective clothing (see above).</td></tr>
</table>

3.3.2 Liquid SO_3 (sulphan) storage and handling

The following safety facilities should be readily accessible in all areas where sulphur trioxide is handled or stored:

- safety showers with quick opening valves which stay open. Water should be supplied through insulated and heat-traced lines to prevent freeze-ups in cold weather;

- eye-wash fountains or other means of washing the eyes with a gentle flow of tap water;

- mild steel storage tanks are used. The tank and all equipment handling liquid SO_3 should be located in a "hot room" maintained at 35 - 42°C. At these temperatures, so close to the liquid SO_3 boiling point of 45°C, pump suction and line pressure drop can cause the liquid to boil or vaporise, in turn causing loss of pumping efficiency and measuring accuracy. The problem can be eliminated if the storage tank is pressurised with dry nitrogen or air (-40°C dewpoint max. temperature), somewhat above the vapour pressure of liquid SO_3 at storage temperatures. Dry air or nitrogen pressure in the storage tank can be used to transfer liquid SO_3 to process by using a DP cell and a flow control valve. Outside the hot room, all lines, valves and equipment containing the liquid must be adequately heat-traced and insulated so that at no time does the surface temperature drop below 35°C. In this way, crystals or polymers will not form and stoppage and plugging problems will be eliminated.

For specific recommendations and technical details about pumps, pipes and valves, the Technical Service Department of the liquid SO_3 supplier should be contacted.

3.3.3 ENGINEERING DATA - PHYSICAL PROPERTY DATA SHEET

NAME: SULPHUR TRIOXIDE		MOLECULAR WEIGHT: 80			
CHEMICAL FORMULA: SO_3 $((SO_3)_2)$					
APPEARANCE:	COLOURLESS GAS; FUMES ON CONTACT WITH ATMOSPHERIC MOISTURE				
SPECIFIC GRAVITY:	2.75 w.r.t. air				
SOLUBILITY (WATER):	g/l	Deliquescent; forms sulphuric acid			
MELTING POINT:	°C	16.8			
BOILING POINT:	°C	44.8			
SPECIFIC HEAT CAPACITY:	kJ/kg°C	0.515			
HEAT OF REACTION TO SO_3 kJ/kg(S)		-3063	-3043	-3018	-2990
$SO_2 + ½ O_2 \rightarrow SO_3$ + HEAT °C		327	427	527	627
EQUILIBRIUM CONSTANT (Kp)	Kp	397	48.1	0.53	2.63
AT TEMPERATURE	°C	400	500	600	700
VAPOUR PRESSURE LIQUID	°C	27 38	44	49	60
SO_3	mmHg	300 600	760	1100	1750

3.4 **65% oleum and 20% oleum**

Oleum is a solution of free SO_3 in 100% sulphuric acid. The commonest commercial strength for oleum is 20%, but 17%, 25% and 30% are also used. The main difficulties in the handling of oleum are:

- its tendency to fume, giving off sulphur trioxide with the amount of fumes being out of all proportion to the size of the liquid source;
- its corrosive effect on many materials of construction;
- its violent, almost explosive reaction with water with the evolution of much heat.

3.4.1 Hazards of oleum 65 and oleum 20

Sulphur trioxide fumes form oleum with atmospheric moisture. The exposure to sulphuric acid mist has been described to some extent in section 3.3.1.: Hazards of sulphur trioxide.

RAW MATERIAL HAZARD DATA SHEET

CHEMICAL NAME	PHYSICAL FORM
Oleum 20 Oleum 65	Viscous liquid

FIRE/EXPLOSION HAZARDS

Non-flammable, explosion risk on contact with water.

REACTIVITY

Reacts violently with water and alkaline materials, SO_3 fumes mixed with air, H_2SO_4 mist.

HEALTH

Fume/mist causes severe irritation if inhaled, causes severe or permanent damage to eyes, severe burns to skin, severe damage to mucous membranes if swallowed.

PROTECTION/HANDLING REQUIREMENTS

FACE/EYES	HANDS	BODY	FEET
Wear full face mask	Wear rubber gloves	Wear overalls and rubber apron	Wear acid/chemical resistant boots

	EMERGENCY ACTIONS			
	INHALATION	EYES	SKIN	INGESTION
FIRST AID	Remove from source of fume/mist Med. att.	Wash with copious amounts of water Med. att.	Wash skin with water Med. att.	Drink 2 glasses water or milk Med. att.
FIRE	Avoid contact with water, keep storage containers cool with water when fighting fires, wear suitable protective clothing (see above) and wear supplied air or self-contained breathing apparatus. In fires gives corrosive fumes/vapour.			

	CONTAIN/ABSORB/ADSORB	FLUSH/WASH AWAY
SPILLAGE LEAK	Treat with an excess of anhydrous sodium sulphate. Wear protective clothing and wear supplied air or self-contained breathing apparatus (see above).	Residual amounts with large volumes of water using a Fog Nozzle. Do not direct jet of water to spillage. Inform authorities.

3.4.2 Oleum storage and handling

Any storage tank must be designed so that branches below the liquid level are kept to a minimum. These will normally be limited to a bottom outlet line and consideration should be given to providing an internal plug valve. An isolation valve should be bolted directly on to the tank branch with an automatic valve in the outlet line.

Automatic valves should be provided with the facility for remote operation from diametrically opposed points and should be operated regularly to prevent sticking. Storage tanks should not be provided with large side manholes because of risk of leakage. Sludge will build up at a rate of 1/4 - 1/2" per year and the design should bear in mind the problem of sludge removal after years of operation. All storage vessels should be fitted with contents level indication and high-level alarm.

All storage vessels must be provided with a system of heating which will maintain the contents at 20 - 25°C (65% oleum). Inside coils should never be used because of the possibility of steam leakage. Outside coils can be used, but should not be in direct contact with the tank surface to avoid the possibility of accelerated corrosion at hot spots. A high standard of maintenance will be required to avoid steam or condensate leaks under the lagging. The upper part of all 65% storage vessels above the maximum liquid level should be heated to about 90°C to prevent

3.4.3 ENGINEERING DATA - PHYSICAL PROPERTY DATA SHEET

NAME	OLEUM 20 (104.5% H_2SO_4) AND OLEUM 65 (114.6% H_2SO_4)	
BOILING POINT	OLEUM 20: 148°C	OLEUM 65: 57°c
VAPOUR PRESSURE		
mm Hg	10 20 35 55	115 200 330 535 760
TEMP.(°C)	50 60 70 80	20 30 40 50 57
FREEZING POINT	CA. -10°C	CA. +0.5°C
VISCOSITY		
Ns/m^2 43°C	16×10^{-3}	18×10^{-3}
SPECIFIC GRAVITY		
kg/m^3	1916	2020
SPECIFIC HEAT		
kJ/kg°C	1.3	1.7
THERMAL CONDUCTIVITY		
W/m°C	0.85	

sublimation of SO_3. This can be achieved by electric tracing or a hot air box. Steam tracing can also be used. All storage tanks must be vented to a scrubber system.

Carbon steel is a satisfactory material for construction of tanks, vessels and pipelines. Stainless steel and special nickel alloys, e.g. Carpenter 20, are very suitable, but expensive and their use is not justified, except where erosion cannot be prevented. Of the non-metallic materials of construction, only PTFE or glass can withstand attack by oleum 65 and 20.

The pipework system should be kept simple with the minimum number of valves installed so that the lines are self-cleaning. Fluid velocity through pipework should be restricted to minimise erosion problems. A velocity of not more than 0.5 m/s is preferable. Sharp bends increase turbulence and should be avoided wherever possible. Cast steel plug valves with fluon sleeve or cast steel ball valves with fluon seal should be used. Centrifugal pumps with special mechanical seals with either reinforced PTFE or ceramic seat are suitable for pumping oleum 65 and 20. Instruments in contact with oleum should normally be fabricated from stainless steel.

For specific recommendation about any technical details, the Technical Service Department of the oleum supplier should be contacted.

3.5 Sulphuric acid and "spent acids"

Sulphuric acid (H_2SO_4) is a dense oily liquid, varying in appearance from water white and clear (from SO_3 absorber) to brown and turbid ("spent acids") according to quality.
"Spent acid" constitutes the sulphuric acid by-product resulting from sulphonation with 20% oleum or SO_3 stripped from 65% oleum:

(i) Sulphonation with 20% oleum

After sulphonation is complete, water is added to the acid products to achieve separation of the alkylbenzene sulphonic acid from sulphuric acid. The "spent acid" represents the lower (high gravity) phase and consists of 78% sulphuric acid, highly coloured material and about 0.1 - 0.2% sulphonic acids. The separation is based on the principle that many sulphonic acids are relatively insoluble in sulphuric acid of intermediate strength (i.e. about 50 - 80%) obtained by suitable dilution of the reaction mixture of sulphonic acid and sulphuric acid after sulphonation with 20% oleum or 100% sulphuric acid. This 70 - 80% sulphuric acid is unlike normal sulphuric acid especially with respect to colour, organic matter and viscosity. It has generally a dark brown/black colour and a strong odour of SO_2.

(ii) Oleum 65-based SO_3/air sulphonation

The SO_3 gas used in this process is generated by thermal stripping from oleum 65. The "spent acid" remaining after the SO_3 gas has been distilled from the 65% oleum is essentially 100% sulphuric acid. This "spent acid" is unlikely to contain sulphonic acids, but it has a dark colour and is unlike a first grade sulphuric acid.

3.5.1 <u>Hazards of sulphuric acid</u>

Sulphuric acid is highly corrosive and causes rapid destruction of body tissue, producing severe burns.

RAW MATERIAL HAZARD DATA SHEET

CHEMICAL NAME	PHYSICAL FORM
Sulphuric acid	Viscous liquid

FIRE/EXPLOSION HAZARDS
Non-flammable.

REACTIVITY
Generates heat when mixed with water. When diluting: add sulphuric acid to water. When heated gives off toxic gases:fumes.

HEALTH
Highly corrosive, causes severe or permanent damage to eyes, severe burns to skin and severe burns to mucous membranes if swallowed.

PROTECTION/HANDLING REQUIREMENTS

FACE/EYES	HANDS	BODY	FEET
Wear full face mask	Wear rubber gloves	Wear overalls and rubber apron	Wear acid/chemical resistant boots

EMERGENCY ACTIONS

	INHALATION	EYES	SKIN	INGESTION
FIRST AID	Remove from source of mist Med. att.	Wash with copious amounts of water Med. att.	Wash skin with water Med. att.	Drink water or milk Med. att.
FIRE	Avoid contact with water, keep storage containers cool with water when fighting fires. Wear suitable protective clothing (see above) and wear supplied air or self-contained breathing apparatus. In fires gives corrosive fumes/vapour.			

	CONTAIN/ADSORB/ADSORB	FLUSH/WASH AWAY
SPILLAGE LEAK	Collect into sealable containers. Wear suitable protective clothing (see above).	Residual amounts with large volume of water. Protective clothing, see above.

Repeated contact of dilute solutions on the skin may cause dermatitis. Concentrated acid will very rapidly cause deep, severe, painful burns. Splashes in the eye rapidly cause severe damage which may be followed by complete loss of sight. Vapour and mists may cause coughing and irritation of the mucous membranes.

Eyewash facilities must always be available and it cannot be stressed too strongly that these should be plentiful in number and kept in prominent and readily accessible position.
The type of eyewash solution recommended is a boric saline solution made up to the following formula:

sodium chloride	9 grams
boric acid	10 grams
distilled water	1 litre

Clean water is an acceptable alternative to boric saline and it must be emphasised that speed of action is absolutely essential with eye injuries and every second may be vital.

The following safety features are required at the intake point:

running water, drain, eyewash, shower, adequate lighting and suitable warning/information notices (Eye Protection Area, No Smoking, hazard data sheets etc.).

3.5.2 Storage and handling

Bulk storage tanks should be sited convenient to the plant and with the intake point in a suitable lay-by in which the delivery vehicle can berth safely. The filling lines and lines to process should be kept as short as possible. Users often prefer to site the tank in an elevated position, so that use can be made of gravity flow to process. The road tanker compressor at its maximum permissible operating pressure of 2 bar g is capable of delivering the acid to a maximum height of 7.5 metres.

Under all conditions of temperature and concentration, sulphuric acid corrodes most metals. The degree of attack on iron and steel is greater with dilute than with concentrated acid. For acid strength of 77% - 85%, carbon steel is recommended and for strengths above 85%, carbon steel or stainless steel can be used for storage vessels.
The slight attack of acid on carbon steel results in the formation of iron sulphate which provides an effective seal against further corrosion unless subjected to weaker acid strengths for prolonged periods.

Pipelines

For 96 - 98% acids, seamless mild steel pipe may be used, but mild steel tube with non-metallic lining, polypropylene or PTFE is preferred. Bends should have a minimum radius of 5 pipe diameters. For continuous duty the velocity should not exceed 0.6 m/s in mild steel pipes.
For about 78% acid, a mild steel pipe with a non-metallic lining, polypropylene or PTFE is preferred. Stainless steels should not be used for acid strengths less than 85%.

Connections between mild steel pipelines should be made by means of bolted flanges. The flanges should be welded internally and externally.

Pumps

A standard end-section centrifugal pump with a special mechanical seal is suitable. The mechanical seal should be enclosed and a compatible flushing liquid provided. Alternatively a glandless pump may be used. Pumps should be protected against running dry.

For specific recommendations about any technical detail, the Technical Service Department of the sulphuric acid supplier should be contacted.

3.5.3 ENGINEERING DATA - PHYSICAL PROPERTY DATA SHEET

NAME:	SULPHURIC ACID (H_2SO_4)				
MOLECULAR WEIGHT 98					
BOILING POINT	°C	186.3	200.6	310	290-317
	% H_2SO_4	76	80	99	100
FREEZING POINT	°C	-22.8	-1.5	5.0	+10.37
	% H_2SO_4	76	88	99	100
78% H_2SO_4 DENSITY					
	kg/m³	1694.4	1684.7	1675.1	1665.7
	TEMP. °C	30	40	50	60
100% H_2SO_4 DENSITY					
	kg/m³	1821.2	1811.7	1801.7	1792.1
	TEMP. °C	30	40	50	60
VISCOSITY		78%	AND	100% H_2SO_4	
	Ns/m²	14.3×10^{-3}	12.8×10^{-3}	24.6×10^{-3}	14.7×10^{-3}
	TEMP. °C	25	40	25	40
THERMAL CONDUCTIVITY SULPHURIC ACID (30°C)	W/m°C	0.52	0.43	0.36	
CONCENTRATION		30	60	90	
SPECIFIC HEAT SULPHURIC ACID (40 °C)	kJ/kg°C	2.0	1.6		
CONCENTRATION %		80	100		

3.6 Alkylbenzenes

Alkylbenzenes are the commonest organic feedstocks found in the detergent industry and, as their name suggests, they consist of a benzene nucleus to which an alkyl chain is bound. The alkyl chain shows a distribution of lengths dependent upon the "cut" of the starting material, while point of attachment to the benzene ring (phenyl-isomer distribution) is governed to a large extent by the manufacturing process (section 3.6.1). The extent of branching of the alkyl chain allows distinction between the two broad categories of alkylbenzenes, "hard" and "soft". Linear alkyl chain compounds, formed from the hydrogen fluoride (HF) catalysed addition of benzene to either a pure olefin, or olefin/paraffin mixture, or from the aluminium chloride ($AlCl_3$) catalysed reaction between chloroparaffins and benzene, are the more biodegradable and are termed biologically "soft". Branched alkyl chain compounds, formed by the $AlCl_3$ catalysed reaction between propylene, condensed to its tetramer, and benzene are less biogradable and are termed biologically "hard".

Linear alkylbenzenes have to a large extent replaced branched alkylbenzenes in the domestic markets because of their improved biodegradability.
No distinction is made between soft alkylbenzenes made from the HF or $AlCl_3$ catalysed processes. However, differences in the physical properties of the neutralised pastes of sulphonic acids, particularly between those derived from hard and soft alkylates (free oil, colour, viscosity), are apparent. Hence each new supply will require testing to determine whether the required properties in the chosen application can be achieved.

3.6.1 Alkylbenzenes: quality

Factories do not, as a general rule, analyse alkylbenzene supplies on a regular basis. Only occasional checks to monitor general quality are recommended, provided there is a long history of supplier reliability. It is however essential to check each batch of alkylbenzene used in research and development programs, or where the supplier is relatively unknown or problems in sulphonation are encountered.

The quality specification of alkylbenzenes is dependent upon the desired performance of the product, which can only be tested under practical conditions. However general guidelines are given below and these may be used to establish a raw material specification with a supplier.

Property	Limits	Comments
Appearance	Colourless, clear liquid	Visual assessment
Sediment	Nil	Visual
Mean molecular weight	± 4 acceptable ± 2 preferable	Branched chain alkylbenzenes by mass spectroscopy
Bromine number	0.05% max.	Suppliers should be encouraged to achieve

		<0.05%. Bromine number detects the presence of unsaturates in the side chain, responsible for colour formation (g Br_2 per 100g)
Moisture	0.05% max.	Karl Fisher
Odour	As agreed	Off-odours may develop on sulphonation. Products should be checked.
Doctor test	Negative	Detects sulphurous components which may lead to off-odours.
Iron content	1 ppm max.as Fe	
Standard sulphonation	98.5%	Sometimes unreliable; analyse free oil
free oil ex standard sulphonation	1.5% max. soft 2.0% max. hard	100% AD basis
2-phenyl isomer	Supplier specification 15-30%	The solubility of alkylbenzene sulphonates increases with decreasing 2-phenyl isomer content. The $AlCl_3$ process generally produces higher 2-phenyl isomer levels (>25%) than the HF process (<20%).
Biodegradability	Legislation	

A proposed specification for LAB to be used in NSD powders could be depicted as follows:

Alkyl chain length distribution:

Less than C10	1% max.
C10	10 - 20%
C10 + C11	40 - 60%
C12	30 - 50%
C13	10 - 25%
More than C13	1 - 10%

LAB	98% min.
Mean molecular weight (MMW)	235-255
Non-sulphonatable material	1.5% max.
Moisture	0.05% max.
Doctor test	Negative
Either Bromine index or	20 mg max.Br_2/100 g
Bromine number	0.05% max.
Density at 15°C	850-880 kg/m^3
Appearance	Colourless clear liquid with no sediment

3.6.2 Alkylbenzenes: safety information

Alkylbenzenes are combustible materials and should be kept away from sources of ignition, e.g. naked flames and sparks. The fire risk is in fact only moderate; alkylbenzene will not readily ignite (flash point typically 120 -150°C) but will burn well if it does. Full respiratory protection is needed when dealing with fires. The vapour is combustible and can explode; all equipment and tanks must be thoroughly cleaned and cleared as safe before welding is permitted.

Alkylbenzenes are only mildly toxic if swallowed, however there is a danger that these low-viscosity hydrocarbon liquids will tend to break up into a foam if the patient vomits and may therefore be aspirated into the lungs - a highly hazardous situation. Effects on the skin and mucous membranes are similar to mineral oils i.e. slightly to moderately irritating, due to their tendency to de-fat the skin. Inhalation of fumes can have a slight narcotic effect, although inhalation is unlikely at normal in-use temperatures.

In the case of skin contact, alkylbenzenes should be washed away with water and then with soapy water. In the case of splashing into the eyes, they should be flushed with copious quantities of clean water and medical attention sought. No specific guidelines are available for exposure limits to alkylbenzene aerosol although by analogy with mineral oils a level of 5 mg/m^3; 8 hour time-weighted average is considered appropriate.

3.6.3 Alkylbenzenes: storage and handling

Bulk storage

Bulk storage of alkylbenzenes is in mild steel tanks maintained at atmospheric pressure by normal venting arrangements. Tanks (firmly anchored) should be sited clear of factory buildings and be bunded to contain materials in the event of leakage. The bund wall should be strong enough to offer complete protection to the tanks in the event of accidental traffic damage. If adequate bund wall strength cannot be provided then the wall must be protected with crash barriers. The bund area should be sized to accommodate the contents of the largest tank plus 10%.

For the types of alkylbenzene used in detergent manufacture, storage tank heating is generally not required (melting point approx. -70°C). Tanks should however be provided with level and temperature indicators (local and remote).

Alkylbenzenes are normally supplied by bulk carrier which may be pressurised or non-pressurised.

Day tanks

For storage at the point of use day tanks may be used although direct connection between main storage tank and metering system is recommended. The principal function of the day tank is to hold a quantity of liquid under constant conditions and thereby prevent process variations. Their capacity should be sufficient for 1 shift plus 20% reserve to allow for possible delays in recharging. The maximum capacity should be limited to 15 m^3 to reduce the load on the building. Day tanks should be installed high enough to provide sufficient head to feed dosing pumps. They should be equipped with temperature indication and level indication/control and high and low-level alarms. The tank overflow should be directed to drains feeding an interception pit.

Direct connection, main storage tank/feed-pump/dosing pump

In figure 4 a flowsheet is presented where the main storage tank is directly connected to the organic dosing pump via a feed-pump. The advantages of such a system are clear:

- no expensive day tanks;
- less dead volume in change over of feedstocks;
- less risk of an empty tank.

Figure 4 Direct connection, main storage tank/feeding pump/dosing pump

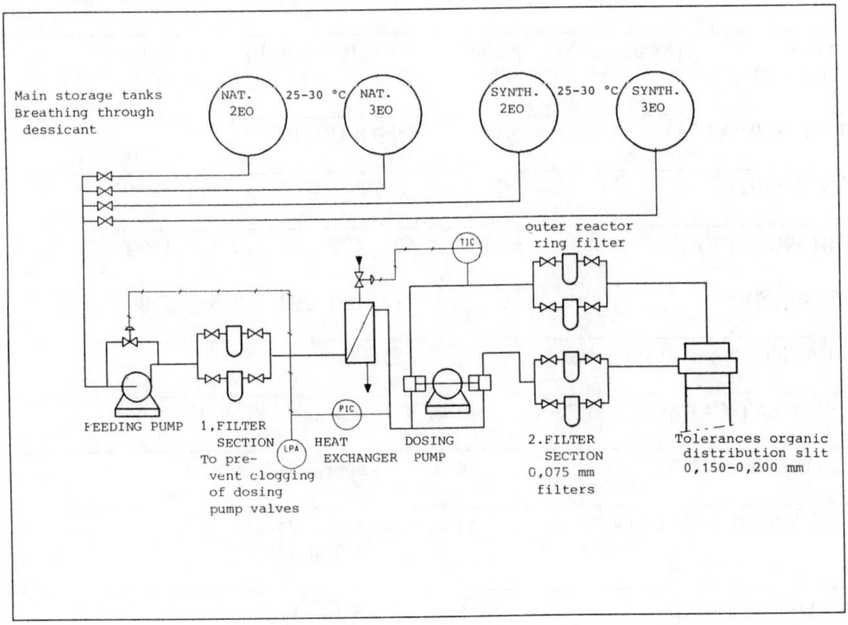

A constant and controlled feed pressure to the dosing pump is required, including a low pressure alarm.
The temperature of the organic feed should be kept constant to maintain constant density and therefore constant mass flow. Filters located immediately upstream of the dosing pump will ensure proper valve operations and a second filtration close to the reactor top will prevent blocking of the organic liquid distribution system.

Pumps, pipes, valves

Pumps and pipework should be of mild steel for alkylbenzenes and valves of cast iron. Centrifugal pumps are recommended for general transfer duties although accurate metering will generally require the use of positive displacement, piston-type metering pumps.

3.6.4 ENGINEERING DATA - PHYSICAL PROPERTY DATA SHEET

Consult supplier for particular feedstocks

NAME: ALKYLBENZENE (ALKYLATE; ALKANE; LAB; BAB; DDB)
MOLECULAR WEIGHT: 240 (variable)

CHEMICAL FORMULA: $C_{18}H_{30}$ (with some distribution)

APPEARANCE: Colourless, sediment free, clear liquid

DENSITY: "Dobane 102"	kg/m^3	859	856		
TEMPERATURE:	°C	15	20		
VISCOSITY: "Dobane 102"	Ns/m^2	10.4x10^{-3}	6.0x10^{-3}	3.4x10^{-3}	
TEMPERATURE:	°C	5	20	40	
SOLUBILITY (WATER):	g/l	INSOLUBLE			
MELTING POINT:	°C	<-70			
VAPOUR PRESSURE:	mbar	52	192	526	975	1000
TEMPERATURE:	°C	217	260	298	325	328
DISTILLATION RANGE AT 1 bar	°C	255-355			
SPECIFIC HEAT CAPACITY at 25°C	kJ/kg°C	1.5-2.0			
HEAT CONDUCTIVITY:	W/m°C	approx 0.1			
HEAT OF VAPOURISATION:	kJ/kg	374 at 25°C 227 at 328°C			
FLASH POINT	°C	>123°C			

3.7 Primary alcohols

Primary alcohols are the second most important class of detergent feedstocks after alkylbenzenes. They are produced either by the catalytic hydrogenation of methyl esters or fatty acids derived from oils and fats e.g. coconut oil or tallow, or from synthetic sources. Alcohols manufactured from Ziegler type processes produce even-numbered chain length primary alcohols. The basic process steps are synthesis of the triethylaluminium catalyst, chain growth from an ethylene precursor, oxidation and finally hydrolysis.

The OXO and Shell Hydroformylation Processes involve the reaction of mixed alpha and internal olefins with hydrogen and carbon monoxide to give predominantly linear primary alcohols, although both processes can yield up to 60% of branched alcohols.

The physical detergency and biodegradation characteristics of primary alcohols are affected by the carbon chain length distribution and therefore each new supply may require testing to determine whether the desired properties in the chosen application can be achieved.

The product of sulphation of detergent alcohols, i.e. an acid sulphate, is unstable and requires immediate neutralisation.

3.7.1 Primary alcohols: quality

Primary alcohols are, to a greater extent than alkylbenzenes, performance rather than specification chemicals. It is the particular characteristic of commercial supplies which are often considered important to the product developer. General guidelines, however, are given below and these may be used to establish a specification with a supplier.

Property	Limits	Comments
Appearance	Colourless clear liquid/ white solid	Visual
Sediment	Nil	Visual
Mean Molecular Weight (MMW)	Specification ±5	-
Water content	0.1% max. preferably 0.05%	High moisture reduces the sulphonation yield and catalyses acid sulphate decomposition reactions by formation of sulphuric acid. Affects production of high active paste.
Saponification value (mg KOH/g)	1.0 max.	Relates to unconverted methyl ester

Acid value (mgKOH/g)	0.2 max.	Relates to unconverted methyl ester
Iodine value (g per 100 g)	0.3 max.	Refers to level of unsaturation and hence potential for coloured by-products.
Carbonyl content (CO)	5 ppm max.	Relates to oxidation
Hydrocarbon	0.5% max.	Normally a paraffinic impurity
Sulphation yield	98% min. C_{12}-C_{14} 95% min. C_{16}-C_{18}	
Biodegradability	Legislation	
Odour	As agreed	Sample

An example of a commercial alcohol from Shell (Dobanol) shows the following physical and chemical properties:

"Dobanol 25" (contains C_{12}, C_{13}, C_{14} and C_{15} primary alcohol) typical values

Mean Molecular Weight (MMW)	207
Hydroxyl no. mg KOH/g	270
Water content % mass	<0.1
Bromine value g Br_2/100 g	0.1
Carbonyl content CO % mass	0.008
Acid value mg KOH/g	<0.1
Density at 25 °C kg/m³	831
Melting range °C	20-22

The C_{16}-C_{18} range fatty alcohol should have the following typical composition:

Molecular weight		250-275
Iodine value		0.5 max
Water content % mass		0.1 max.
Iron content		1 ppm max. as Fe
Hydroxyl number (mg KOH/g)		205-220
Chain length distribution		
	C_{14}	0- 2%
	C_{16}	40-50%
	C_{18}	40-50%
	C_{20}	0- 3%

3.7.2 Primary alcohols: safety information

Primary alcohols are combustible materials and should be kept away from sources of ignition. The fire risk is considered moderate for detergent type alcohols with flash-points in the range 107-170 °C. The vapour is combustible and can explode, therefore all equipment and tanks must be thoroughly cleaned and cleared as safe before welding is permitted.

In the unlikely event of accidental ingestion of primary alcohols, the patient should be given water to drink but vomiting should not be induced.

At normal operating temperatures the vapour pressure of primary alcohols is insufficient to pose an inhalation hazard. Alcohols should be washed from the skin with soap and water. In the event of splashing into the eyes, they should be washed with copious quantities of clean water and medical attention sought. When handling these materials, goggles and rubber or plastic gloves should be worn.

No specific guidelines are available for exposure limits to alcohol aerosol although by analogy with mineral oil a level of 5 mg/m^3; 8 hour time-weighted average is considered appropriate.

3.7.3 Primary alcohols: storage and handling

Bulk storage

The bulk storage of primary alcohols is similar to that required for alkylbenzenes (section 3.6.3.) with the following differences:

- primary alcohols can be solids or liquids. Consequently it is necessary to maintain the storage tank at a temperature of 5 °C above the melting point. Hot water (temperature dependent upon application) must be used to heat the storage tanks and for pipeline tracing. The pipe carrying the hot water to the heating coil in the storage tank should enter through the top of the tank so that as material melts around the pipe, a passage is created for venting any pressure which might develop during a re-melting process;

- it is recommended that stainless steel (316) or glass reinforced plastic (GRP) is used, although mild steel tanks can be used for some primary alcohol raw materials providing that tank surfaces are free from rust particles and that nitrogen blanketing is used to prevent contact with atmosphere and rust formation.

Day tanks/direct connection system

Day tanks or a direct system similar to that specified for alkylbenzenes are recommended fabricated in the appropriate materials of construction (SS316).
Day tanks should incorporate hot water heating coils.

Pumps, pipes and valves

Pumps, pipework and valves should be of stainless steel. Pump types are similar to those for alkylbenzenes.

3.7.4 ENGINEERING DATA - PHYSICAL PROPERTY DATA SHEET

Consult supplier for particular feedstocks

NAME:	PRIMARY ALCOHOL MOLECULAR WEIGHT: LAUREX LI : 202 AVERAGE ALFOL 16-18: 253 AVERAGE				
CHEMICAL FORMULA: $C_nH_{2n+1}OH$					
APPEARANCE:	LAUREX LI (C_{12}-C_{14} Natural) - very soft solid ALFOL 16-18 (C_{16}-C_{18} Synthetic) - white solid				
DENSITY:	LAUREX LI ALFOL 16-18	kg/m^3	830 814		
TEMPERATURE:	LAUREX LI ALFOL 16-18	°C	25 60		
VISCOSITY:	LAUREX LI	Ns/m^2	$12X10^{-3}$		
TEMPERATURE:		°C	37		
SOLUBILITY (WATER):		g/l	INSOLUBLE		
MELTING POINT:	LAUREX LI ALFOL 16-18	°C	25-27 46-49		
BOILING POINT:	LAUREX LI ALFOL 16-18	°C	265 270-355		
VAPOUR PRESSURE: mbar LAUREX LI	0.007	0.053	0.3	0.63	14.34
TEMPERATURE: °C	40	60	80	90	140
SPECIFIC HEAT CAPACITY: LAUREX LI	kJ/kg°C	2.0 at 37°C			
THERMAL CONDUCTIVITY: LAUREX LI	W/m°C	appr. 0.15 (calc.) at 37°C			
HEAT OF VAPOURISATION: LAUREX LI	kJ/kg	appr. 489 at 24°C			
FLASH POINT: LAUREX LI ALFOL 16-18	°C	appr. 130 appr. 160			

3.8 Primary alcohol ethoxylates

Primary alcohol ethoxylates are an important class of detergent feedstock for anionic active manufacture. They are made by the addition of ethylene oxide to a primary alcohol in the presence of an alkaline catalyst. The addition of the second ethylene oxide molecule to the alcohol is kinetically favoured in comparison with the addition of the first ethylene oxide, hence the product of ethoxylation contains a distribution of ethylene oxide chain lengths attached to the alcohol along with the starting alcohol itself. Consequently the physical, detergency and biodegradation characteristics are affected not only by the carbon chain length distribution as is the case for primary alcohols, but also by the ethylene oxide distribution which in turn can be supplier dependent.

The commonest alcohol ethoxylates found as feedstocks for sulphation have an average of 2 to 3 molecules of ethylene oxide (2EO or 3EO).

During the sulphation of alcohol ethoxylates the by-product 1,4-dioxane may be formed. Although the formation of 1,4-dioxane is governed predominantly by the sulphation and neutralisation conditions and by the chemical composition of the feedstock, other factors such as the quality of the raw material also contribute. These factors must be considered during the storage and handling of the alcohol ethoxylate feedstock.

3.8.1 Primary alcohol ethoxylate quality

The formation of the by-product 1.4 dioxane during the sulphation of alcohol ethoxylates has prompted a re-appraisal of the required quality standards for this feedstock. Consequently some of the material components which are considered to affect 1.4 dioxane formation should be included when establishing a specification with a supplier. The following information may be used as guideline:

Property	Limits	Comments
Appearance	White to pale straw liquids & solids	Visual
Sediment	Nil	Visual
Mean Molecular Weight	Specification ±5	-
Water content	0.1% max. 0.05% recom- mended	Moisture forms sulphuric acid during sulphation which is responsible for hydrolysis and considered to catalyse 1,4-dioxane formation

Odour	As agreed	-
Polyethylene glycol (PEG)	0.5% max.	Reduces biodegradability, surface-active properties and increases 1,4-dioxane
Free ethylene oxide	1 ppm max.	International quality specification safety limit
Iron	2 ppm max.	-
Cloud point	Specification ±1°C	-
Biodegradability	Legislation	-

It should be noted that the LESAC committee concluded that the formation of allergens was associated with the quality of the feedstock and ethylene oxide used as well as the processing conditions. It is imperative that guarantees are obtained from suppliers of ethoxylated alcohols that they will adhere strictly to the conditions specified as follows:

a) the olefin content of the starting alcohol must be carefully controlled at the lowest possible level which should not be higher than 0.9%;

b) the ethylene oxide used should contain no more than 40 ppm acetaldehyde and an absolute maximum of 60 ppm non-volatile residue. The desired maximum non-volatile residue is 30 ppm. Ethylene oxide made from ethylene chlorohydrin must not be used;

c) manufacture of the alcohol 3 EO intermediate will be under conditions such that the temperature of ethoxylation does not exceed 180°C.

Two examples of commercial primary alcohol ethoxylates from Shell ("Dobanol" ethoxylates) show the following physical and chemical properties:

<u>"Dobanol 23 - 2 EO" and "Dobanol 25 - 3 EO" typical values</u>

	"Dobanol 23-2"	"Dobanol 25-3"
MMW	282	339
% Mass EO (calc.)	31.2	39.0
HLB value (calc.)	6.2	7.8
Water content % mass	<0.1	<0.1
Ash content % mass	<0.3	<0.3
PEG content % mass	3.0 max.	3.0 max.
Acid value mg KOH/g	0.05-0.2	0.05-0.2
Density kg/m³ (20°C)	910	930
Melting range °C	4-5	5-6

3.8.2 Primary alcohol ethoxylates: safety information

Alcohol ethoxylates generally possess similar properties to their parent primary alcohols and therefore the same precautions and procedures as outlined in section 3.7.2 should be followed.

In addition to the precautions outlined for primary alcohols, rubber or plastic boots and aprons should be worn if there is a possibility of substantial body contact.

3.8.3 Primary alcohol ethoxylates: storage and handling

Similar storage and handling procedures to those outlined for primary alcohols (section 3.7.3) are recommended with the following exceptions:

Bulk storage

The bulk storage of primary alcohol ethoxylates is similar to that required for alkylbenzenes (section 3.6.3) with the following differences:

- primary alcohol ethoxylates can be solids or liquids. Consequently it is necessary to maintain the storage tank at a temperature of 5 °C above the melting point. Hot water (temperature dependent upon application) must be used to heat the storage tanks and for pipeline tracing. The pipe carrying the hot water to the heating coil in the storage tank should enter through the top of the tank so that as material melts around the pipe, a passage is created for venting any pressure which might develop during a re-heating process;

- primary alcohol ethoxylates are hygroscopic and therefore, in order to maintain the specified low moisture content of the feedstock up until the point of use, it is recommended that storage tanks (and day vessels if used) be vented through desiccant (silica gel);

- it is recommended that stainless steel (316) or glass reinforced plastic (GRP) is used for the storage of primary alcohol ethoxylates.

Day tanks/direct system

Day tanks or direct systems similar to those specified for alkylbenzenes are recommended, to be fabricated in the appropriate materials of construction (SS316). Day tanks should incorporate hot water heating coils and be provided with a desiccant venting system.

Pumps, pipes and valves

Pumps, pipework and valves should be of stainless steel. Pump types are similar to those for alkylbenzenes.

3.8.4 ENGINEERING DATA - PHYSICAL PROPERTY DATA SHEET

Consult supplier for particular feedstocks.

NAME:	PRIMARY ALCOHOL ETHOXYLATES MOLECULAR WEIGHT: "Dobanol 23-3EO":325		
CHEMICAL FORMULA:	C_nH_{2n+1} $(CH_2CH_2O)_mOH$		
APPEARANCE:	Liquids and solids		
DENSITY:	kg/m^3	913	
TEMPERATURE:	$^\circ C$	40	
VISCOSITY:	Ns/m^2	$21x10^{-3}$	$9x10^{-3}$
TEMPERATURE	$^\circ C$	30	60
SOLUBILITY (WATER):	g/l	INSOLUBLE	
MELTING POINT:	$^\circ C$	4-6	
FLASH POINT (PMCC):	$^\circ C$	154	
FREE ALCOHOL:	wt%	20.0 max.	
ACID VALUE:	mg KOH/g	0.05 - 0.20	
SPECIFIC HEAT CAPACITY:	$kJ/kg^\circ C$	1.9 (calc.)	
THERMAL CONDUCTIVITY:	$W/m\ ^\circ C$	0.1 (calc.)	

3.9 Alpha-olefins

Alpha-olefins are a potential replacement for alkylbenzenes in detergent applications. Their exploitation, however, is largely limited to the Far East, centred on Japan, at present. Commercial supplies of alpha-olefins are produced by the oligomerisation of ethylene with the main exponents being Shell, Gulf and Ethyl. The material contains quantities of internal and vinylidene olefins but only when the level of internal olefins becomes excessive is this likely to have an effect on feedstock conversion.

The physical, detergency and biodegradation characteristics of alpha-olefins are affected by the carbon chain length distribution and therefore each new supply may require testing to determine whether the desired properties for the new chosen application can be achieved.

The neutralised product of alpha-olefin sulphonation requires hydrolysis to remove sultones which are skin sensitisers.

Approximate typical C_{14}-C_{16} olefin composition of Shell, Gulf and Ethyl can be summarised as follows:

Component	Shell	Gulf	Ethyl
Linear alpha-olefin	95%	91%	76%
Vinylidene	2	7	19
Linear internal	3	1	5
Di-olefins	<0.05	<0.05	<0.05
Paraffin	0.1	1.5	0.4

3.9.1 Alpha-olefins: quality

When developing a specification for the supply of alpha-olefins the following guidelines may be applied, although it should be remembered that the distribution of alpha, internal and vinylidene olefins are supplier-specific depending upon the manufacturing process:

Property	Limits	Comments
Appearance	Colourless liquids, white solids	-
Sediment	Nil	-
Mean Molecular Weight	Specification ±5	-
Moisture content	0.01 wt% max.	Can produce highly coloured species via formation of sulphuric acid
Carbonyl content mg/kg	10 max.	Reduction in yield
Peroxide as active oxygen mg/kg	5 max.	Potential fire/explosion hazard
Alpha-olefin	60% min.	Supplier specific
Total mono olefins	99.0% min.	-
Paraffins	0.2 wt% max.	Reduction in yield

Two examples of commercial linear alpha-olefins show the following physical and chemical properties:

SHOP (Shell) C_{16} Linear Alpha-Olefin (1-hexadecene)

Property	Typical figures
Appearance at 20°C	Clear liquid
Odour	Mild
MMW	225
Density at 20°C kg/m^3	780
Freezing point °C	3
Viscosity at 20°C Ns/m^2	3.1 x 10^{-3}
Distillation range °C	
IBP	270
5%	280
50%	282
90%	290
Water mg/kg	<100
Carbonyl C=O mg/kg	< 15
Peroxide as active O mg/kg*	< 3
Carbon distribution % mass (gas chromatography)	
C_{14}	< 4
C_{16}	> 92
C_{18}	< 4
Hydrocarbon type % mass (gas chromatography)	
n alpha-olefins	> 92.0
Branched olefins (mainly vinylidenes)	< 4.0 } <8.0
Beta internal olefins	< 2.5 }
Paraffins	< 0.2

* Subject to addition of anti-oxidant.

Gulftene (Gulf) C_{16} Linear Alpha-Olefins (1-hexadecene)

Property	Typical figures
Appearance at 20°C	Clear liquid
Density at 20°C kg/m^3	785
Freezing point °C	4
Distillation range °C	
5%	270
95%	300
Carbon distribution % mass C_{16}	95 min.
Hydrocarbon type % mass	
n alpha-olefins	91 min.
Mono olefins	98.5 min.
Paraffins	1.5 max.

3.9.2 Alpha-olefins: safety information

Generally alpha-olefins can be considered as similar to ethoxylated alcohols and similar precautions should be taken (section 3.8.2). However on exposure to air, alpha-olefins will gradually form organic peroxides which constitute a fire and explosion hazard.

3.9.3 Alpha-olefins: storage and handling

In terms of storage and handling alpha-olefins should be considered as similar to alkylbenzenes (section 3.8.3.) with the following differences:

- alpha-olefins may be solid or liquid. Consequently it is necessary to maintain the storage tank at a temperature of 5°C above the melting point, e.g. C_{16} linear alpha-olefin has a freezing point of 3°C. Hot water (temperature dependent upon application) must be used to heat the storage tanks and for pipeline tracing. The pipe carrying the hot water to the heating coil in the storage tank should enter through the top of the tank so that as material melts around the pipe, a passage is created for venting any pressure which might develop during a re-heating process;

- alpha-olefins are prone to oxidation and hence nitrogen blanketing is recommended.

3.9.4 ENGINEERING DATA - PHYSICAL PROPERTY DATA SHEET

Consult supplier for particular feedstocks.

NAME:		ALPHA-OLEFINS (C_{14}-C_{18}) MOLECULAR WEIGHT: 225 (C_{16})			
CHEMICAL FORMULA:		C_n-H_{2n}			
APPEARANCE:		Oily liquid			
DENSITY:		kg/m³	781		
TEMPERATURE:		°C	20		
SOLUBILITY (WATER):		g/l	INSOLUBLE		
MELTING POINT:		°C	4		
VISCOSITY AT 20°C		Ns/m²	3.1×10^{-3}		
BOILING POINT:		°C	285		
VAPOUR PRESSURE: mbar	105.3	263.2	526.3	789.5	1000
TEMPERATURE: °C	200.7	230.8	257.1	274.2	284.4
SPECIFIC HEAT CAPACITY:		kJ/kg°C	1.34 at 25°C 1.66 at Bpt		

HEAT CONDUCTIVITY:	W/m °C	0.14 (calc.)
HEAT OF VAPOURISATION:	kJ/kg	282.1 at 25°C 177.6 at 127°C
FLASH POINT	°C	127

3.10 Fatty acid methyl esters (FAME)

FAME may become in the future a possible organic feedstock to be sulphonated to Fatty Acid Methyl Ester Sulphonate (FAMES). This feedstock is naturally renewable as it is produced from oils/fats or fatty acids. There are several possible process routes for the manufacture of FAME. Transesterification of fat triglycerides is the predominant method for manufacture of mixed fatty acid methyl esters, and direct esterification of fatty acids (FA) is practised if very selective cuts of product, in general as an intermediate detergent range alcohol, are desired. Methyl cocoate is a mobile, oily liquid above 25°C with a yellow tint and a characteristic fatty acid pungent odour. FAME sulphonation to FAMES is technically possible but hardly applied up to now (1990).

3.10.1 FAME quality

One of the most important requirements to produce quality methyl ester sulphonate is the use of a high quality methyl ester feedstock. The best of the feedstocks currently used are highly refined palm oil derivatives. These feedstocks have low acid values , a narrow molecular weight distribution, contain little or no paraffinic compounds and have been extensively hydrogenated to reduce their iodine value to a very low value (less than 0.5). The latter is essential because high iodine values indicate the presence of double bonds in the feedstock which will result in dark coloured products upon sulphonation.

Property	Hydrogenated palm kernel fatty acid Me-ester	Hydrogenated tallow fatty acid Me-ester
C_{10}	1.2	-
C_{12}	59.1	0.7
C_{14}	21.9	4.0
C_{16}	8.8	29.8
C_{18}	9.0	64.6
C_{20}	-	0.9
Average chain length	C13.1	C17.6
Acid value	0.2	0.2
Sapon. number	240	197
Hydroxyl number	2.4	0
Moisture content %	<0.1	<0.1
Iodine value	<0.5	<0.5
Melting range °C	30-35	40-45
MMW (Mean MW)	233	284

3.10.2 FAME: storage and handling

Bulk storage

Fatty acid methyl esters will readily hydrolyse if stored in contact with water, the hydrolysis being acid catalysed. Therefore steps should be taken to exclude water from the storage vessel. It is therefore recommended that storage tanks be vented through desiccant (silica gel).

Bulk storage of fatty acid methyl esters is in stainless steel 316 tanks maintained at atmospheric pressure by normal venting arrangements. Tanks should be sited clear of factory buildings and be bunded to contain materials in the event of leakage. The bund area should be sized to accommodate the contents of the largest tank plus 10%. The melting temperature of the tallow based Me-ester will be about 35 °C, consequently it is necessary to maintain the storage tank at a temperature of 5 °C above the melting point.

Hot water (temperature dependent upon application) should be used to heat the storage tanks and for pipeline tracing. The pipe carrying the hot water to the heating coil in the storage tank should enter through the top of the tank so that as material melts around the pipe, a passage is created for venting any pressure which might develop during a re-heating process.

Day tanks/direct systems

Day tanks or direct systems similar to those specified for alkylbenzenes are recommended. Day tanks should incorporate hot water heating coils.

Pumps, pipes and valves

Pumps and pipework should be of mild steel, valves of cast iron. Centrifugal pumps are recommended for general transfer duties.

3.10.3 ENGINEERING DATA - PHYSICAL PROPERTY DATA SHEET

NAME:	FATTY ACID METHYL ESTER METHYL COCOATE	
CHEMICAL FORMULA:	$C_nH_{2n+1}COOH$ MOLECULAR WEIGHT: 236	
APPEARANCE:	OILY LIQUID	
DENSITY: TEMPERATURE:	kg/m^3 $^\circ C$	870 25
MELTING POINT:	$^\circ C$	7-10
BOILING POINT:	$^\circ C$	245-262
VISCOSITY AT 50°C:	Ns/m^2	2.1×10^{-3}
SPECIFIC HEAT:	$kJ/kg^\circ C$	2.0
HEAT CONDUCTIVITY:	$W/m^\circ C$	0.16 (calc.)

3.11 **Sodium hydroxide (caustic soda) and Sodium hydroxide solution (45-50%) (caustic soda liquor)**

Caustic soda is made by the electrolysis of brine using both mercury cells and diaphragm cells to give caustic soda of different qualities.

Mercury cell caustic soda liquor is a high-grade product, low in chloride and chlorate impurities and, because of its traditional associations, known as Rayon grade. It is produced directly by amalgam decomposition at the strength at which it is sold. This usually contains 46.8% NaOH (i.e. 100° Tw caustic liquor).

Diaphragm cells yield a weak caustic liquor which also contains unelectrolysed salt. This undergoes further processing which involves evaporation and salt removal to yield caustic soda liquor of a strength of 46.8%, known as ordinary grade. This grade contains about 1% sodium chloride together with a small quantity of chlorate.
Caustic soda as a liquor is preferred because delivery can be made in bulk, handling costs are low and no redissolving equipment is required. The overall cost per ton of caustic soda is lowest with the liquid form. Commercial caustic soda liquor is marketed at a strength of about 47%, corresponding to a eutectic point of +6°C on the freezing point curve. Transport and storage difficulties due to freezing in cold weather are thus minimised.

Fused and flake anhydrous caustic are also available, mainly for export. Flake is particularly convenient when using caustic soda in small quantities.

3.11.1 Sodium hydroxide quality

General quality specifications for caustic soda and caustic soda liquor are given below and these may be used to establish a raw material specification with a supplier.

Sodium hyroxide solid

Property	Limits	
	Mercury cell	Diaphragm cell
Appearance	Off-white	Off-white
NaOH	Min. 98	Min. 95
Sodium carbonate %	Max. 2	Max. 3
Water insolubles %	Max. 0.05	Max. 0.05
Sodium sulphate %	Max. 0.1	Max. 0.1
Sodium chloride %	Max. 0.05	Max. 2.0
Sodium chlorate mg/kg	Max. 100	Max. 1600
Iron as Fe mg/kg	Max. 10	Max. 10
Copper as Cu mg/kg	Max. 5	Max. 5
Arsenic as As mg/kg	Max. 2	Max. 2
Mercury as Hg mg/kg	Max. 2	Max. 2

Sodium hyroxide solution (45-50%)

Property		Limits	
		Mercury cell	Diaphragm cell
Appearance		Colourless	Colourless
NaOH		Min. 45	Min. 45
Sodium carbonate %		Max. 1	Max. 1
Water insolubles %		Max. 0.3	Max. 0.3
Sodium sulphate %		Max. 0.05	Max. 0.05
Sodium chloride %		Max. 0.03	Max. 1.0
Sodium chlorate	mg/kg	Max. 50	Max. 800
Iron as Fe	mg/kg	Max. 5	Max. 5
Copper as Cu	mg/kg	Max. 1	Max. 2
Arsenic as As	mg/kg	Max. 1	Max. 1
Mercury as Hg	mg/kg	Max. 1	Max. 1

3.11.2 Sodium hydroxide safety information

Caustic soda destroys organic tissues. Contact with the eyes rapidly causes severe tissue damage, and on the skin it causes burns often accompanied by deep ulceration with ultimate scarring. Goggles, gloves and protective clothing must be worn by persons involved with caustic soda handling. Supplies of running water must be available. In the event of contact with skin or eyes, wash copiously with water for a long period and obtain medical attention.

RAW MATERIAL HAZARD DATA SHEET

CHEMICAL NAME	PHYSICAL FORM
Caustic soda (NaOH)	Colourless solid flake

FIRE/EXPLOSION HAZARDS
Non-flammable.

REACTIVITY
Can ignite combustible materials, generates heat when mixed with water.

HEALTH
Causes severe or permanent damage to eyes, severe burns to skin and severe burns to mucous membrane if swallowed.

PROTECTION/HANDLING REQUIREMENTS

FACE/EYES	HANDS	BODY	FEET
Wear chemical workers goggles or full face mask	Wear rubber gloves	Wear overalls	Wear rubber boots

	EMERGENCY ACTIONS			
	INHALATION	EYES	SKIN	INGESTION
FIRST AID	Remove from source of dust Med. att.	Wash with copious amount of water Med. att.	Wash with water Med. att.	Drink water or milk Med. att.
FIRE	Material non-flammable, when fighting fires involving this material do not use water Jet or water spray extinguisher. Protective clothing, see above.			
	CONTAIN/ABSORB/ADSORB		FLUSH/WASH AWAY	
SPILLAGE LEAK	Sweep up and collect spillage into suitable sealable container. Protective clothing, see above.		Residual amounts carefully with water, neutralise with dilute acid, flush away with large volumes water. Clothing see above.	

RAW MATERIAL HAZARD DATA SHEET

CHEMICAL NAME	PHYSICAL FORM
Caustic soda (NaOH, 45% sol.)	Solution of appr. 45% NaOH

FIRE/EXPLOSION
Non-flammable.

REACTIVITY
Generates heat when mixed with water.

HEALTH
Causes severe or permanent damage to eyes, severe burns to mucous membrane if swallowed.

PROTECTION/HANDLING REQUIREMENTS

FACE/EYES	HANDS	BODY	FEET
Wear chemical workers goggles or wear full face mask	Wear rubber gloves	Wear overalls and rubber apron	Wear rubber boots

	EMERGENCY ACTIONS			
	INHALATION	EYES	SKIN	INGESTION
FIRST AID	Remove from source of mist Med. att.	Wash with copious amounts of water Med. att.	Wash with water Med. att.	Drink water or milk Med. att.
FIRE	Material non-flammable, when fighting fires involving this material do not use water Jet or water spray extinguisher. Protective clothing, see above.			
	CONTAIN/ABSORB/ADSORB	FLUSH/WASH AWAY		
SPILLAGE LEAK	Collect spillage into suitable container, protective clothing see above and mist/ dust mask.	Residual amounts carefully with water, neutralise with acid, flush away with large volume of water.		

3.11.3 Storage and handling

Storage

47-50% caustic soda liquor should be stored at normal (20°C) or slightly elevated temperatures in mild steel tanks of welded construction. If the liquor temperature is allowed to reach 40°C or more for liquors containing 30% or more caustic, or reach 60°C or more for lower concentrations, tanks must be stress-relieved. Except for the vent and overflow connections, storage tanks should be closed vessels to avoid carbon dioxide pick-up from the air which would produce a sodium carbonate sludge in the liquor.

A small amount of sediment forms during storage and off-take pipes from tanks should be arranged to draw liquor from a point about 0.1 m above the base. Soda liquor should normally be allowed to settle for 24 hours prior to use. Periodic cleaning of storage tanks to remove sediment may be necessary; when this is done, special arrangements should be made for disposal of the sediment, which may have a high mercury content, to approved waste treatment agencies or the soda liquor supplier. The liquor will freeze in winter and it is advisable for outside tanks to be heated and for pipelines to be lagged. Storage vessels can be heated by internal steam coils. Internal heating elements or steam coils which are fabricated from carbon steel must be stress-relieved.

For steam heating coils, austenitic stainless steel, e.g. AISI 316, is recommended in preference to mild steel. It is better practice to have a larger surface area of heating coil at a lower temperature than vice versa, as a means of avoiding stress cracking at elevated temperatures. The temperatures above which 304 and 316 type stainless steel are susceptible to stress corrosion in caustic liquors vary with concentration between 115 and 240°C. L.P. steam should therefore be used.

Alternatively, a steam coil can be fitted outside a tank beneath its base. External coils must be separated from the base of a tank by an air gap of at least 3" and should be supported from below to avoid local hot spots due to heat transfer through brackets to the tank itself. The coil should be enclosed within a heating chamber to exclude draught. Tanks sited outdoors will require thermal insulation on surfaces external to the heating chamber.

For neutralisation of sulphonic acids, dilute caustic soda solutions are required to obtain AD pastes in the range of 45-70% AD. Therefore the necessity of heated outdoor storage can be avoided by diluting caustic liquor with cold water at the time of delivery. Dilution to 27% w/w NaOH will lower the freezing point of the solution to -7°C.

The addition of water to a strong caustic soda solution is accompanied by the evolution of heat. This effect may cause local boiling of the liquor accompanied by violent spurting. If it is not possible to mix in a stirred tank, the addition is best made by directing the stream of water along the side of the tank to give a stirring motion to the liquor. The combined volumes of the liquor and the dilution water must be turned over at least twice by a recirculation pump to ensure formation of a homogeneous solution free from layering.

It is recommended to bund caustic soda liquor tanks. The bund should be able to retain 110% of the contents of the tank and should be constructed from a strong cement with a low water-to-cement ratio and should be lined with a caustic-resistant material if the temperature of the stored caustic is above ambient. Suitable materials are bitumen, furane-based resin screed or one of the "plastics" PVC, polypropylene or polyethylene. Bunds and catchment areas should drain to an effluent treatment system for neutralisation before final discharge to drain.

Pipework and valves

Carbon steel seamless piping with welded-on flanges is recommended for the construction of liquor mains. Pipelines, normally 2" nominal bore, should be self-draining, but if this is impracticable, drain cocks should be installed at the lowest points. Where there is a risk of liquor freezing, pipelines should be steam-traced and lagged.

Pumps

For transferring caustic soda liquor from a road or rail tanker to storage or from storage to the point of use centrifugal pumps are recommended. Pumps should be all ferrous fitted, e.g. cast-iron casing, stainless or cast-iron impeller, stainless steel or mild steel shaft. Pump glands or seals should be covered by an anti-splash guard and the pump should be housed over a mild steel catchment tray which drains to a sump. The wetted end of pumps sited outdoors should be heated and lagged (LP steam spacer tracing may be used).

Alternatively the pump can be housed in a box which encloses an LP steam coil positioned so that it does not touch the pump, to avoid local hot spots. All pumps should have facilities for emptying the pump chamber and washing it out with cold water.

3.11.4 ENGINEERING DATA - PHYSICAL PROPERTY DATA SHEET

Name: Sodium Hydroxide liquor
Chemical formula: NaOH MW40

Density (15 °C) kg/m^3

% NaOH by weight	Density kg/m^3
10	1112
15	1167
20	1223
25	1278
30	1332
40	1434
45	1483
50	1530

Freezing point (°C) of caustic soda solutions

% NaOH by weight	Commencement of freezing [°C]
10	-10
15	-22
20	-25
25	-14
30	+ 1
40	+15
45	+ 6
50	+12

Boiling point caustic soda solution °C

% NaOH by weight	Boiling point °C
10	111
20	114
30	118
40	124
50	142

Viscosity Ns/m^2

%NaOH by weight	Viscosity ($x10^{+3}$)		
	20°C	30°C	40°C
10	2.0	1.5	1.0
20	4.5	3.5	2.0
30	15	9	4
40	50	20	8
50	80	40	15

Specific heat kJ/kg°C

%NaOH by weight	Specific heat kJ/kg°C (at 25°C)
10	3.86
20	3.65
30	3.55
40	3.45
50	3.24

Thermal conductivity W/m°C

%NaOH by weight	Thermal conductivity (25°C)
10	0.63
20	0.64
30	0.64
40	0.65
50	0.65

3.12 Sodium carbonate (soda ash)

Sodium carbonate is used as an alkaline builder in NSD powders and in a few locations as a neutralising agent for alkylbenzene sulphonic acids and as a chemical buffering agent during caustic soda neutralisation. It is commercially available in two forms, a light and dense powder. For the neutralisation process a 20% solution is prepared on site by dissolving sodium carbonate in water.

3.12.1 Sodium carbonate quality

A guideline for a specification is given below:

Property	Limits
Appearance	White crystalline
Sodium bicarbonate	Min. 98%
Water	Max. 1%
Water insolubles	Max. 0.5%
Sodium bi-carbonate	Max. 1%
Iron as Fe	Max. 10 ppm
Copper as Cu	Max. 1 ppm
Arsenic as As	Max. 10 ppm

3.12.2 Safety information

Excessive exposure to carbonate dust can cause temporary irritation of the skin, the mucous membranes and the eyes. Excessive inhalation of dust can cause coughing and shortness of breath. Goggles should be worn where dust is prevalent. In the event of accidental contact with the eyes, wash out with copious quantities of water. A sodium carbonate solution does not cause dermatitis but if the skin is in contact with the solution for a prolonged period the natural oils are removed and the skin may become dry and cracked, allowing infections to occur from other sources. It is important, therefore, in such cases to replace the natural oils by frequent use of a cream made by mixing equal parts of lanoline and vaseline or petroleum jelly.

RAW MATERIAL HAZARD DATA SHEET

CHEMICAL NAME	PHYSICAL FORM
Sodium Carbonate (Na_2CO_3)	White powder

FIRE/EXPLOSION HAZARDS
Non-flammable

REACTIVITY
Reacts violently with acidic materials.

HEALTH
Can cause severe irritation to eyes. Dust causes irritation if inhaled.

PROTECTION/HANDLING REQUIREMENTS

FACE/EYES	HANDS	BODY	FEET
Wear chemical workers goggles	Wear gloves	Wear overalls	Wear normal factory safety footwear

	EMERGENCY ACTIONS			
	INHALTATION	EYES	SKIN	INGESTION
FIRST AID	Remove from source of dust	Wash with copious amount of water Med. att.	Wash with water	Drink water or milk Med. att.
FIRE	Material non-flammable When fighting fires involving this material do not use water jet or water spray extinguisher. Protective clothing, see above and mist/dust mask.			

	CONTAIN/ABSORB/ADSORB	FLUSH/WASH AWAY
SPILLAGE LEAK	Sweep up/collect spillage into a suitable container. Protective clothing, see above and mist/dust mask.	Residual amounts, flush away with large volumes of water.

3.12.3 Storage and handling

Anhydrous sodium carbonate is hygroscopic and, as well as absorbing moisture, will combine with atmospheric CO_2 to form bicarbonate. Ageing in this way leads to the formation of agglomerate lumps and crystals. It is essential, therefore, that the material is stored under cover in dry conditions. The material should be stored away from acids, with which it reacts vigorously forming carbon dioxide gas. Conical bottoms of silos should have a slope of not less than 60°.

For the neutralisation process Na_2CO_3 is dissolved batch-wise in a mild steel tank which is provided with an agitator and recirculation. The anhydrous Na_2CO_3 is added to 40°C warm water, and agitation and recirculation is continued until a clear solution has been obtained. The final concentration of the unsaturated solution (see diagram showing the solutibility of Na_2CO_3 in water) is roughly in the 20-25% range, depending on final AD paste concentration (40-45% AD).

3.12.4 ENGINEERING DATA - PHYSICAL PROPERTY DATA SHEET

Name:	Sodium carbonate
Chemical formula:	Na_2CO_3 MW 106

Hydrates (see also solubility diagram - figure 5)

$Na_2CO_3.10H_2O$, stable below 32°C
$Na_2CO_3.7H_2O$, 32-35°C
$Na_2CO_3.1H_2O$, 35-100°C

Solubility in 100 g water at:

25°C	21.4 g Na_2CO_3
50°C	32.0 g Na_2CO_3

Heat of solution in water at 18°C

+23.4 kJ/gmole Na_2CO_3
+ 9.2 kJ/gmole $Na_2CO_3.1H_2O$
-87.4 kJ/gmole $Na_2CO_3.7H_2O$
-68.1 kJ/gmole $Na_2CO_3.10H_2O$

Density of Na_2CO_3 solutions at 40°C (kg/m³)

Na_2CO_3 conc. in water (w/w %)	Density kg/m³
15%	1150
20%	1200
25%	1220

Viscosity of Na_2CO_3 solutions at 20°C, 30°C and 40°C

% Na_2CO_3 by weight	Viscosity NS/m^2		
	20°C	30°C	40°C
15	2.5×10^{-3}	1.9×10^{-3}	1.5×10^{-3}
20	-	3.0×10^{-3}	2.3×10^{-3}
25	-	4.8×10^{-3}	3.5×10^{-3}

Specific heat (kJ/kg°C) of Na_2CO_3 solutions at 50°C

%Na_2CO_3 by weight	Cp kJ/kg°C
10	3.9
20	3.7
25	3.5

Figure 5 Sodium carbonate solubility in water

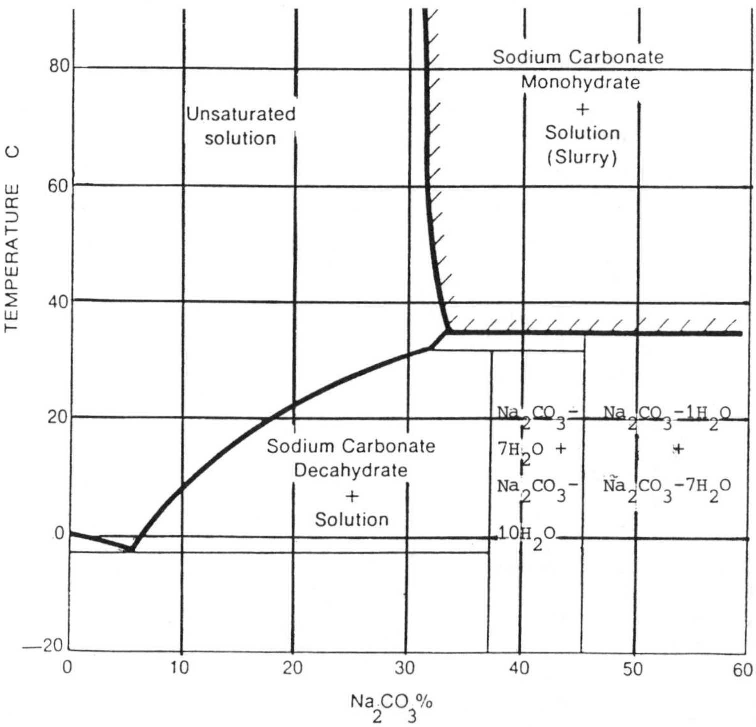

3.13 Ammonium hydroxide (ammonia liquor)

Ammonia is produced by the Haber-Bosch process. A hydrogen/nitrogen mixture, obtained from steam treatment of natural gas, is purified and passed over metallic oxide catalyst under pressure. The gases react at temperatures above 450°C forming ammonia gas which is absorbed in water to form ammonium hydroxide.

3.13.1 Ammonium hydroxide quality (NH_4OH)

Ammonium hydroxide is used for the neutralisation of alkyl hydrogen sulphates ($R\text{-}O\text{-}SO_3H$) and alkyl ether sulphates ($R\text{-}(OCH_2CH_2)_nOSO_3H$).

Guidelines for specification:

Property	Limits
Odour	Pungent and typical
Ammonia content	Min. 25%
Specific gravity	Max. 0.91
Non volatiles mg/kg	Max. 20
Iron as Fe mg/kg	Max. 2

3.13.2 Safety information

The pungent odour of ammonia provides ample warning of its presence. The gas is irritating to the eyes and mucous membranes of the respiratory tract. The maximum allowable exposure limits for ammonia vapour are:

- 18 mg/m^3 (25 ppm); 8 hour time-weighted average;
- 27 mg/m^3 (35 ppm); 10 minute time-weighted average.

The liquor can irritate the skin and can cause corneal ulcers if splashed into the eyes. Persons handling ammonia liquor must wear PVC gloves and gas-tight goggles. Eye-wash bottles and breathing apparatus, preferably of the self-contained type, must be available on site. In case of contact with the skin, wash off with water. If ammonia liquor splashes into the eyes, wash out with copious volumes of water or saline solution.

RAW MATERIAL HAZARD DATA SHEET

CHEMICAL NAME	PHYSICAL FORM
Ammonium hydroxide (NH_4 OH)	Colourless solution of appr. 35% in water

FIRE/EXPLOSION HAZARDS
Non-flammable.

REACTIVITY
Can cause explosion when mixed with many chemicals. When heated gives toxic fume (NH_3 gas).

HEALTH
Can cause severe irritation to eyes, skin and irritation to mucous membranes if swallowed.

PROTECTION/HANDLING REQUIREMENTS

FACE/EYES	HANDS	BODY	FEET
Wear full face mask	Wear rubber glo- ves	Wear overalls and rubber apron or wear overalls and PVC apron	Wear normal fac- tory safety foot- wear

EMERGENCY ACTIONS

	INHALATION	EYES	SKIN	INGESTION
FIRST AID	Remove from source of mist vapour Med. att.	Wash with co- pious amounts of water Med. att.	wash with water	Drink milk. Med. att.

FIRE	Cut off source of material from fire, keep storage containers cool with water. Protective clothing, see above.

	CONTAIN/ADSORB/ABSORB	FLUSH/WASH AWAY
SPILLAGE LEAK	Collect into suitable sealable containers, protective clothing see above and wear chemical cartridge respirator.	Residual amounts, neutralise with dilute acid, flush away with large volumes of water.

3.13.3 Storage and handling

In Europe ammonium hydroxide is supplied in winter and summer grades i.e. the solution strength is altered so as to minimize the vapour pressure corresponding to varying ambient temperatures. For winter conditions it is supplied as 34-35% w/w ammonia and during the summer the concentration is reduced to 32-33% w/w so that a maximum vapour pressure of 1.4 to 1.7 bar is maintained. For tropical conditions ammonium hydroxide should be obtained at a lower concentration, 25% w/w, which would give approximate corresponding vapour pressures of 1.65 bar at 50°C, 1.4 bar at 45°C and 1.2 bar at 40°C.

Storage

Copper and its alloys, nickel, zinc, galvanised surfaces and aluminium alloys are all attacked by ammonia. Iron, steel, pure aluminium and stainless steel are suitable for the construction of storage vessels. Ammonia liquor is conveniently handled in mild steel. Bulk storage installations for ammonia liquor must be bunded and at a reasonable distance (depending on size of installation) from dwelling houses and places of public assembly. Ammonia liquor tanks should be bunded by themselves and never in a common bund with any other substance with which ammonia will react. Pumps should be sited within the bund, and the filling connection should terminate just inside the bund wall. Ammonia liquor should be stored in closed tanks to prevent ammonia escaping to the atmosphere and to prevent the absorption of atmospheric carbon dioxide.

Since the liquor is capable of developing modest vapour pressures the storage tank must be fitted with a suitable relief valve, set in accordance with the design working pressure, either piped to free atmosphere or to an ammonia scrubber.

In general, vapour vent lines must be led to a height and location where they are remote from access by personnel and from where any vapour discharge is carried clear of all buildings and populated areas. Equally, since the vessel pressure can fluctuate below as well as above atmospheric pressure, a vacuum relief valve should be fitted.

Off-loading bulk road tankers are usually 18 tons general purpose tankers which are normally discharged by compressed air from the tanker's own compressor at up to 2-2.4 bar absolute. On completion of transfer, this gas pressure will immediately be felt in the storage tanks. For compressed air discharge it is recommended that tanks be designed to 2.4 bar absolute and tested to 3 bar absolute. Storage tanks designed to accept compressed air discharge from tankers must always be splash filled. The ideal capacity for a storage installation is two weeks use plus one tanker load (18 tons).

Pipework and valves

The most convenient piping for ammonia liquor systems is standard mild steel water pipe. This may be used with screwed, butt-welded or glanded joints as required. Valves may be of steel or cast-iron but must not contain brass or bronze fittings. The most satisfactory valves are the "Saunders" diaphragm valve or any of the high quality ball valves. Polyethylene or polypropylene plastic piping is quite satisfactory within the specified temperature and pressure limits. Compression fittings containing copper and brass must not be used.

Pumps

Pumps for handling ammonia liquor must have all the "wetted parts" made from iron, mild or stainless steel. Pumps must always be sited below the tank from which they are taking suction, and the inlet piping should be as short as possible and generously sized to minimise pressure drop. Where vapour locking is likely to be an unavoidable problem, positive displacement pumps must be used. Valves situated in pump inlet systems should have a low pressure drop.

3.13.4 ENGINEERING DATA PHYSICAL PROPERTY DATA SHEET

NAME:	Ammonium hydroxide solution
CHEMICAL FORMULA:	NH_4OH
CONCENTRATION % NH_3 w/w:	35-36% in winter 32-33% in summer
DENSITY (15 $^\circ$C):	kg/m^3 870-880 for 35-36% sol. 880-890 for 32-33% sol.
FREEZING POINT $^\circ$C	-94 for 35-36% sol. -100 for 32-33% sol.
ATMOSPHERIC BOILING POINT $^\circ$C	18 for 35% sol. 20 for 32% sol.
HEAT OF DILUTION	1.74 kJ/gmole NH_3 dilution from about 25% to 0.5% w/w

VAPOUR PRESSURE 30% LIQUID	mbar	313	763	1723
TEMPERATURE	$^\circ$C	0	20	40

3.14 **Chlorine (Cl_2)**

The major source of chlorine is from electrolysis processes as a by-product of the electrolytic manufacture of caustic soda or by electrolysis of fused sodium chloride or magnesium solution. Chlorine gas is dried by cooling and passing through concentrated sulphuric acid before being liquefied by compression at low temperature. Chlorine gas may be used to prepare "in-situ" diluted NaOCl solutions for bleaching of AD pastes.

3.14.1 Chlorine quality

Property	Limits
Appearance	Clear yellow liquid giving off greenish yellow gas
Chlorine	Min. 99.5% by vol.
Residue on evaporation % w/w	Max. 0.02
Water content % w/w	Max. 0.01
Nitrogen trichloride ppm	Max. 10

3.14.2 Chlorine safety information

Chlorine is a non-flammable gas. At low concentrations the gas is a respiratory irritant. At high concentrations and/or long exposures it causes irreversible damage to lung tissue. Chlorine gas is so intensely irritating that concentrations of 10-15 mg/m^3 in air are readily detectable by the normal person. The maximum allowable exposure limits of chlorine gas are:

- 3 mg/m^3 (1 ppm); 8 hour time-weighted average;
- 9 mg/m^3 (3 ppm); 10 minutes time-weighted average.

The gas is 2.5 times heavier than air. The gas or liquid will react readily with many organic chemicals - in some cases with explosive violence e.g. with hydrocarbons, alcohols and ethers. Chlorine reacts with ammonia or ammonium compounds to form mixtures of chloramines. One of these might be the explosive nitrogen trichloride.

The volume of liquid chlorine increases considerably with increasing temperature. For example, in a cylinder with 75 kg, roughly corresponding to a volume of 52 litres, the volume will increase 2.1 litres when the temperature rises from 0°C to 20°C, i.e. the volume expands 4% over 20°C, a coefficient of cubic expansion of 0.002 per $^\circ$C.

A suitable respirator should be worn when opening container valves or breaking chlorine connections. The Puretha canister type respirator is usually adequate for coping with a gas leak from a chlorine container valve. For the more hazardous situations like equipment cleaning and repair, PVC suits and protective headgear are recommended. Self-contained breathing apparatus must be available for use if required. Emergency showers and eye fountains must be provided. Chlorine containers and piping should be hydrostatically tested to 20 bar and thoroughly dried before introducing chlorine. Ventilation of plant areas must provide for a one to four minute rate of air change and the suction of the fans must be located at or near floor level.

To find a leak, tie a cloth to the end of a stick, soak the cloth with ammonia liquor and hold close to the suspected area. A white cloud of ammonium chloride will indicate a chlorine leak. Never use water on a chlorine leak. The corrosive action will make the leak worse and the heat evolved will cause liquid chlorine to evaporate faster.

RAW MATERIAL HAZARD DATA SHEET

CHEMICAL NAME	PHYSICAL FORM
Chlorine (Cl_2)	Greenish, yellow gas

FIRE/EXPLOSION HAZARDS
Non-flammable, oxidising agent

REACTIVITY
Reacts violently with organic materials, metals, phosphoros, can produce toxic gases/fumes on contact with water.

HEALTH
Toxic if inhaled and causes severe irritation to eyes and to skin.

PROTECTION/HANDLING REQUIREMENTS

FACE/EYES	HANDS	BODY	FEET
Wear full face mask or wear chemical workers goggles	Wear PVC gloves and cotton liners	Wear overalls and PVC apron	Wear normal factory safety footwear

	EMERGENCY ACTIONS			
	INHALATION	EYES	SKIN	INGESTION
FIRST AID	Remove from source of fumes Med. att.	Wash with copious amounts of water Med. att.	Wash with water	Not applicable
FIRE	Cut off source of material from fire, keep storage containers cool with water. Protective clothing, see above, including supplied air or self contained breathing apparatus.			

	CONTAIN/ABSORB/ADSORB	FLUSH/WASH AWAY
SPILLAGE LEAK	Isolate source of leakage. Wear suitable protective clothing (see above) and respiratory protection.	Not applicable.

3.14.3 Storage and handling

Dry chlorine, in either the gaseous or liquid state, can be handled satisfactorily in carbon steel containers/cylinders. The dry chlorine is transported in cylinders up to 75 kg capacity or in one-ton containers. All containers, whether empty or full, must be stored in a dry area and protected from external heat sources such as steam pipes. If stored outdoors, they must be protected from direct sunlight, rain and snow. Chlorine containers should be segregated from other compressed gas containers. Containers must never be stored near anhydrous ammonia, hydrocarbons or flammable materials. The storage area must be well ventilated and underground storage must be avoided.

Cylinders should be stored in an upright position and secured so that there is no risk of falling. They should not be stacked or racked more than one high. Storage should be arranged so that the cylinders are used in the order in which they are received, and so that individual cylinders can be removed and daily inspection can be carried out without disturbing other cylinders. Full and empty cylinders should be stored separately.

The discharge of gas from a container is accompanied by cooling of the contents. When gas is withdrawn, the temperature and pressure of the chlorine will fall to a fairly constant level depending on the amount of heat which can push through the body of the container. This governs the subsequent rate of evolution of the gas. At atmospheric temperatures of about 15°C there is a steady and continuous flow of about 1.4 kg of gas per hour from a 75 kg cylinder and between 4.5 and 9 kg from a 900 kg container. Quantities greater than these can be obtained by connecting two or more cylinders or drums in parallel.

Discharge rates may be increased by circulating air at room temperature (20-25°C) around the cylinders with a fan. Never place the container in a bath of hot water or apply direct heat (high coefficient of cubic expansion of liquid chlorine and rise in pressure with temperature i.e. at 0°C - 3.5 bar, 20°C - 6.7 bar, 40°C - 11 bar).

A flexible connection should be used between container/ cylinder and the piping system. Copper tubing suitable for 35 bar is recommended. A vacuum breaking device should be used to prevent suck-back of liquid when the ton-container becomes empty. No attempt must be made to repair chlorine equipment while it is in operation and when piping systems are in service.
When a chlorine system is to be cleaned or repaired, tank, pipelines and other equipment must always be purged with dry air or nitrogen as a safeguard to health. This is especially important where cutting or welding operations are undertaken, since iron and steel will ignite in chlorine at about 230-260°C. When propylene pipe is used to carry chlorine gas through a manned production area, the plastic pipe should be supported throughout its length in metallic tubing. The metallic pipe should be clearly marked so that it is easy to identify that it carries chlorine gas. As with all other hazardous raw materials, written safety procedures (and where applicable, a "permit to work certificate" system) must be established.

3.14.4 ENGINEERING DATA - PHYSICAL PROPERTY DATA SHEET

NAME:	Chlorine (liquid and gas)		
CHEMICAL FORMULA:	Cl_2 M.W. 71		
		Gas	Liquid
DENSITY at 0°C -760 mmHg kg/m³		3.2	1468
20°C		2.5	1410
COEFF. OF CUBIC EXPANSION			0.0021 per °C
BOILING POINT:	760 mm Hg °C		-34
VAPOUR PRESSURE	0°C (bars abs.)		3.5
	20°C (bars abs.)		6.7
	40°C (bars abs.)		11.0
VISCOSITY	Ns/m² (20°C)		0.5×10^{-3}
SPECIFIC HEAT	(kJ/kg°C)	0.47	1.0
HEAT OF VAPOURISATION	(0°C) kJ/kg		265.5
	(-34°C) kJ/kg		288.5
HEAT OF REACTION OF CHLORINE WITH SODIUM HYDROXYDE LIQUOR	kJ/kg of Cl_2	1462	
SOLUBILITY OF CHLORINE IN WATER kg chlorine per m³ of water	10°C		9.97
	15°C		8.5
	20°C		7.29
	25°C		6.41

3.15 Sodium hypochlorite

Sodium hypochlorite is most commonly produced by a continuous process in which dilute caustic soda solution and chlorine are fed into a specially designed, cooled reactor. Control of rate of chlorine addition is by monitoring of Redox potential. Temperature is controlled to below 38°C to avoid chlorate formation. Alternately, a batch process may be used in which chlorine is passed slowly into a cold 20°C NaOH solution through a sparge tube near the bottom of a tank. Air agitation is used and chlorine feed rate is kept low to minimise temperature rise.

Finished solution is settled to remove metallic impurities and thus improve stability. Sodium hypochlorite solutions are normally supplied at 14-15% available chlorine wt% which roughly corresponds with the sodium hypochlorite concentration, as the molecular weight of NaOCl (74.5) is almost equal to the MW of Cl_2 (71.0).

Reaction equations:

$$2NaOH + nH_2O + Cl_2 \rightarrow NaOCl + NaCl + (n+1)H_2O$$

3.15.1 Sodium hypochlorite quality

As supplied, sodium hypochlorite solution meets the following specification:

Property	Limits
Appearance	Clear liquid
Available chlorine % wt	14.6
Sodium hypochlorite % wt	15.3
Sodium hydroxide % wt	0.4
Water insolubles ppm	Max. 10
Iron as Fe ppm	Max. 2
Gassing rate (37 °C) ml/day	Max. 55

Cloudiness in hypochlorite solution can sometimes be indicative of a greater propensity for gassing. The exact causes of variations in gassing rate are not yet fully understood although it is known that the presence of metals, e.g. Fe, Co, Ni and Cu encourage gassing. Gaseous decomposition follows the simple mechanism:

$$2NaOCl \rightarrow 2\,NaCl + O_2$$

3.15.2 Hypochlorite safety information

Sodium hypochlorite is highly corrosive causing damage to the skin, mucous membranes and eyes. All personnel handling or using hypochlorite should wear chemical type goggles, overalls which are resistant to chemicals, plastic or rubber gloves and rubber boots.

The same protective clothing should be worn for dealing with spillages but with face shields instead of goggles. Gas masks must be available in areas where hypochlorite is handled. Trouser legs should be worn outside the rubber boots.

A flexible supply of clean running water must be available where tankers are off-loaded and where hypochlorite is handled or used. Provision of an emergency shower is essential.

RAW MATERIAL HAZARD DATA SHEET	
CHEMICAL NAME Sodium hypochlorite (NaOCl)	PHYSICAL FORM Yellow green solution 12-15% active chlorine

FIRE/EXPLOSION HAZARDS
Non-flammable, oxidising agent, explosion risk with many chemicals.

REACTIVITY
Reacts violently with combustible materials, reducing agents, acidic materials, decomposes on heating.

HEALTH
Causes severe or permanent damage to eyes and burns to skin.

PROTECTION/HANDLING REQUIREMENTS			
FACE/EYES	HANDS	BODY	FEET
Wear chemical workers goggles or full face mask	Wear rubber gloves	Wear acid/ chemical resistant over-alls	Wear rubber boots

	EMERGENCY ACTIONS			
	INHALATION	EYES	SKIN	INGESTION
FIRST AID	Remove from source of mist Med. att.	Wash with copious amount of water Med. att.	Wash with water Med. att.	Drink water or milk Med. att.
FIRE	Material non-flammable, keep storage containers cool with water, when fighting fires involving this material wear supplied air or self contained breathing apparatus.			

	CONTAIN/ABSORB/ADSORB	FLUSH/WASH AWAY
SPILLAGE LEAK	Collect spillage in a suitable sealable container. Protective clothing, see above and chemical cartridge respirator.	Residual amounts. Flush away with large volumes of water.

3.15.3 Storage and handling of NaOCl solutions

Stability of hypochlorite solutions

Sodium hypochlorite solutions decompose slowly on standing with the production of sodium chloride, oxygen and sodium chlorate:

$2NaOCl \rightarrow NaCl + NaClO_2$ (slow)
$NaOCl + NaClO_2 \rightarrow NaCl + NaClO_3$ (fast)
$2NaOCl \rightarrow NaCl + O_2$ (gaseous decomposition)

The nature and rate of decomposition are affected by a number of factors listed below:

(i) pH of solution

Dilute hypochlorite solutions are most stable at pH 13 and least stable at pH 6.7. For this reason sodium hypochlorite solutions are normally stabilised by the addition of free NaOH solution.

(ii) Temperature

As a rough approximation a solution of 10% NaOCl stored at 15-20°C would be expected to lose 0.5-1% available chlorine in a period of a month, assuming a pH above 11.5. If, however, the solution is stored at higher temperatures the av. Cl_2 of the solution may fall to 5% in one month. Consequently hypochlorite bleach solution is much more stable at low temperatures. The rate of decomposition may double for each increase in temperature of 5°C, a factor that makes cold storage desirable.

(iii) Metals

Problems due to liberation of oxygen gas ("gassing") arise when the hypochlorite is subject to catalysis by trace metal impurities. Co, Ni and Cu greatly enhance the formation of oxygen while iron does so to a lesser extent. The gassing rate is much less at lower temperatures and at lower av. Cl_2 strengths.

(iv) Effect of initial hypochlorite concentration (and temperature of storage) on the rate of decomposition of NaOCl solutions is shown below.

Available Cl_2 in original solution	Half-life days at	
wt %	25°C	15°C
16.0	44	175
8.9	220	800
4.6	790	5000

In higher temperature climates a lower available Cl_2 level may have to be accepted due to the lower stability at higher temperatures.

Storage

Storage tanks should be constructed from glassfibre reinforced polyester (GRP) with an unplasticised PVC lining and have a minimum size of 25m^3. This is based on 20-25% in excess of a normal tanker delivery. Vertical cylindrical tanks are recommended with a sump and a drain point to allow complete emptying before cleaning. To assist the cleaning operations which must be carried out at 6-monthly intervals, at least one manhole should be provided. As a minimum procedure the tank must be washed out with water and ventilated by opening the manholes, then atmosphere checked with a Standard Draeger air tester for chlorine and mercury (ex caustic soda) vapours.

The tanks are filled by means of an external PVC pipe from the delivery tanker to the top of the tank. It is recommended that tankers built to withstand pressure be discharged by means of air pressure from a compressor or from an independent clean, dry, oil-free air supply. Where tankers cannot withstand pressure they should be emptied by means of a titanium centrifugal pump. A drain point should be installed on the intake pipe to allow the pipe to be emptied into a container before breaking the connection from the tanker.

The storage tank must be vented through an alkaline absorbing liquid or charcoal filters and a 100mm diameter pipe leading from the top of the tank to the bund wall. This eliminates the possibility of chlorine escaping during tank filling. It also acts as an overflow. The normal outlet from the tank is raised 10-15cm from the bottom to prevent entrained particulate matter from entering the process. Prior to cleaning, the tank is emptied by means of a bottom valve on the sump. A bund wall around all the hypochlorite stock tanks is necessary, with a maximum capacity of one stock tank plus 10%. The bund can be drained by means of an external plastic pump.

Where hypochlorite storage in sunny positions cannot be avoided or high ambient temperatures are prevalent, provision must be made for maintaining the outer surface of the tanks at a low temperature and thereby ensuring a low hypochlorite temperature.

The hypochlorite temperature should preferably be not more than 20°C and not higher than 25°C under any circumstances. The recommended cooling system is an in-line titanium heat exchanger which may be used to cool incoming hypochlorite or in a recirculation mode to keep the contents of the bulk storage tank cool.

Pipework and pumps

Pipework is in PVC and hypochlorite is transferred through titanium centrifugal pumps (wetted parts: solid titanium 30; seals: double mechanical seal arranged for water flush supply between seals.

3.15.4 ENGINEERING DATA - PHYSICAL PROPERTY DATA SHEET

NAME:	Sodium hypochlorite
CHEMICAL FORMULA:	NaOCl M.W. 74.5
DENSITY AT 15°C, 14.5% av.Cl_2	1240-1280

SOLUBILITY DATA - wt %	NaOCl solubility
15°C	30.6
20°C	34.8
25°C	46.0

3.16 Hydrogen peroxide

Hydrogen peroxide may be used as a bleaching agent in fatty acid methyl ester sulphonation.

3.16.1 Hydrogen peroxide quality

Hydrogen peroxide is inherently very stable but the presence of minute amounts (less than ppm) of some catalytic impurities (ferrous/ferric ions, cuprous and cupric ions) can give rise to quite high rates of decomposition. The pH of the solution can affect both the solubility of a catalytic ion as well as its activity as a decomposition agent.

In the presence of iron, for instance, the rate of decomposition increases markedly between pH 3 and 4, due to the formation of very active colloidal hydrated iron oxide. At pH greater than 5 most catalytic ions are almost insoluble and because of this the pH is normally adjusted to 4.5-6.0. The pH also influences the rate of decomposition of uncontaminated solutions of hydrogen peroxide. In alkaline solution the rate of decomposition increases rapidly as the pH is increased.

3.16.2 Hydrogen peroxide safety information

Concentrated solutions of hydrogen peroxide (i.e. about 30%) are strong irritants of the skin and respiratory passages. Contact causes a whitening of the skin which will return to its normal colour after prolonged washing with water. Longer contact causes blistering.

The maximum allowable exposure limits of H_2O_2 vapour are:

- 1.5 mg/m^3 (1 ppm); 8 hour time-weighted average exposure limit;
- 3 mg/m^3 (2 ppm); 10 minute time-weighted average exposure limit.

Keep in a cool place away from sources of heat. Avoid contamination. Wear suitable protective clothing, gloves and eye/face protection.

RAW MATERIAL HAZARD DATA SHEET

CHEMICAL NAME	PHYSICAL FORM
Hydrogen peroxide (H_2O_2)	Colourless solution 32% H_2O_2 in water

FIRE/EXPLOSION HAZARDS
Non-flammable, oxidising agent.

REACTIVITY
Reacts violently with many organic materials, decomposes violently.

HEALTH
Causes severe or permanent damage to eyes, burns to skin, mist causes severe irritation.

PROTECTION/HANDLING REQUIREMENTS

FACE/EYES	HANDS	BODY	FEET
Wear chemical workers goggles or full face mask	Wear rubber glo-ves	Wear overall and rubber apron	Wear rubber boots

EMERGENCY ACTIONS

	INHALATION	EYES	SKIN	INGESTION
FIRST AID	Remove from source of mist Med. att.	Wash with co-pious amounts of water Med. att.	Wash skin with water	Drink water or milk Med. att.
FIRE	Material non-flammable, keep storage containers cool with water. Protective clothing, see above and dust/mist mask.			

	CONTAIN/ABSORB/ADSORB	FLUSH/WASH AWAY
SPILLAGE LEAK	Do not use absorbents or ad-sorbents, must wash/flush away.	Do not allow to evaporate, wash away immediately with large volumes of water. Clo-thing etc., see above.

3.16.3 Storage and handling

Aluminium of 99.5% minimum purity is the preferred material for storage vessels and pipelines. Hydrogen peroxide, as delivered, will not significantly corrode aluminium over long periods, and the corrosion products do not seriously affect the stability of the chemical. However, in the presence of chloride ions, serious pitting can occur, so dilution water must be of acceptable purity and contamination with chlorides must be avoided.

Fully austenitic stainless steel may be used in service with H_2O_2, including storage. The preferred grade of stainless steel for storage and transport vessels is AISI 304 L.

Plastics are generally less satisfactory as constructional materials for hydrogen peroxide than aluminium or stainless steel. However, provided that due allowances are made for these shortcomings, certain plastics can be used for the fabrication of small installations. The two plastics approved are high density polyethylene and the system of rigid PVC reinforced with glass fibre/polyester resins. It must be noted that other common plastics such as polypropylene, glass reinforced polyesters alone and nylon are unsuitable for long term service with H_2O_2. They are subject to unpredictable embrittlement and "ageing".

Piping from plastic tanks may be in aluminium, stainless steel or rigid PVC. Metallic pipework should be used where there exists any serious risk of physical damage. Some fluorinated plastics and rubbers can be used in service with H_2O_2, for example, polytetrafluorethylene, polyvinylidene fluoride, and a co-polymer of vinylidene fluoride and hexafluoropropylene such as VITON.

The recommended material of construction for pumps is stainless steel. The use of pumps with packed glands that require lubricants is not recommended but instead use of pumps with mechanical seals with wetted parts made from ceramic/glass-filled polytetrafluorethylene. Ball valves should be used, with suitable arrangements for releasing pressure of gas built up from decomposition of H_2O_2 trapped in the valve. Thus, acceptable valves have a seat, usually polytetrafluorethylene, which lifts at a predetermined pressure or, alternatively, a small hole drilled in the ball so that in the "off"position the channel through the ball is in communication with the upstream liquid side.

3.16.4 ENGINEERING DATA - PHYSICAL PROPERTY DATA SHEET

NAME:	HYDROGEN PEROXIDE			
CHEMICAL FORMULA:	H_2O_2	MOLECULAR WEIGHT: 34.0		
APPEARANCE:	CLEAR WATER WHITE SOLUTION			
DENSITY:	kg/m^3	1101	1131	1243
CONCENTRATION:	w/w %	27.5	35	60
VISCOSITY:	20°C, appr. 30% strength, NS/m^2, 1.1 x 10^{-3}			

MELTING POINT:	°C	-22.5		-32.5	
CONCENTRATION:	w/w %	27.5		35	
VAPOUR PRESSURE (30%)	mbar	1450	3160	6320	11800
TEMPERATURE:	°C	20	30	40	50
SPECIFIC HEAT:	kJ/kg°C (30%)		3.6		
THERMAL CONDUCTIVITY:	W/m °C		approx. 0.55		
HEAT OF DILUTION:	kJ/kg (30% w/w)		3.0		

3.17 **Ethyl alcohol (Ethanol)**

Industrial ethyl alcohol may be added to lauryl ether sulphates (LES) in order to reduce the viscosity. Ethyl alcohol is produced on a large scale by the hydration of ethylene which is derived from petroleum cracking processes.

3.17.1 Ethyl alcohol quality

Ethyl alcohol meets the following specification:

Property	Limits
Appearance	Colourless liquid
Acidity as H_2SO_4, %	Max. 0.005
Residue on evaporation, %	Max. 0.01
Aldehydes and sultones as acetaldehyde, %	Max. 0.2

3.17.2 Ethyl alcohol safety information

The vapour does not present an appreciable health hazard under normal conditions of use, although prolonged exposure to high concentrations (above 5000 ppm) may produce respiratory irritation, headache, dizziness or narcosis. The maximum allowable exposure limits for ethyl alcohol vapour are:

- 1900 mg/m^3 (1000 ppm);8 hour time-weighted average exposure limit.

Repeated contact of the liquid with the skin can have a degreasing effect which makes it more susceptible to infection.

RAW MATERIAL HAZARD DATA SHEET

CHEMICAL NAME	PHYSICAL FORM
Ethyl alcohol (C_2H_5OH)	Colourless liquid

FIRE/EXPLOSION HAZARDS
Highly flammable, may produce flammable mixtures on contact with air, distant ignition possible.

REACTIVITY
Reacts violently with oxidising agents.

HEALTH
Causes irritation to eyes, vapour can cause irritation to eyes and to mucous membranes if inhaled.

PROTECTION/HANDLING REQUIREMENTS

FACE/EYES	HANDS	BODY	FEET
Wear chemical workers goggles	Wear rubber gloves	Wear overalls	Wear rubber boots

EMERGENCY ACTIONS

	INHALATION	EYES	SKIN	INGESTION
FIRST AID	Remove from source Med. att.	Wash with copious amount of water Med. att.	Wash with water	If large amounts are swallowed Med. att.
FIRE	Fight fires with drypowder or alcohol resistant foam extinguisher or carbon dioxide or water spray. Keep storage containers cool with water.			

	CONTAIN/ABSORB/ADSORB	FLUSH/WASH AWAY
SPILLAGE LEAK	Collect into suitable sealable container, residual amounts absorb in sand or inert absorbent, protective clothing see above.	Flush area of spillage with large volumes of water.

3.17.3 Storage and handling

Storage

Ethanol is not corrosive to the common construction metals except aluminium and may be stored or processed satisfactorily in iron, mild steel or copper vessels. Ethanol is a highly flammable liquid whose vapours form ignitable and explosive mixtures with air. Storage tanks should be in the open surrounded by bund walls. They should be closed vessels vented via flame arresters into an area free from sources of ignition. All pipelines should be sloped so as to drain back into the tank or to a suitable drain point. For offloading a rigid inlet delivery line with a hose connection appr. 1m above the ground must be provided. The delivery vehicle must be effectively earthed before discharging.

A permanent earthing point should be installed as close as possible to the tanker connecting point.
The earthing cable should terminate with a spring clip which should be clamped to the tanker barrel. The filling line should be fitted with a ball- or gate valve. The total length of hose and charging lines should not exceed 30m. The storage tank can be fitted with external gauge glasses of the reflex type. Reflex type gauge glasses are virtually unbreakable and therefore as safe as is reasonably practicable. To measure the quantity of solvent leaving the tank, an integrating meter, can be installed. Discharge pipes into containers must be arranged to discharge below liquid level.

Pumps

To transfer solvent a centrifugal pump is recommended. This should be a self-priming pump if the level of liquid falls below the level of the pump suction. Only metal pumps and pipework may be used and must be earthed.

The selection of electrical apparatus is related to the assignment of risk, e.g. Zone 0, 1 or 2 as specified in International Electrotechnical Recommendation IEC 79-10. The Zone determines the method of explosion protection required for the electrical equipment, and this affects the cost of installation.

3.17.4 ENGINEERING DATE - PHYSICAL PROPERTY DATA SHEET

NAME:	Ethyl alcohol (Ethanol)
CHEMICAL FORMULA:	CH_3CH_2OH (M.W. 46)
DENSITY 20°C kg/m³:	790
MELTING POINT °C:	-110.5
BOILING POINT °C	78.3
VAPOUR PRESSURE AT 20°C mm Hg	43

VISCOSITY AT 20°C Ns/m^2	$1.2x10^{-3}$
EXPLOSIVE LIMITS IN AIR % v/v	3.3-19.0
AUTO-IGNITION TEMPERATURE °C	390-439
SPECIFIC HEAT AT 20°C kJ/kg°C	2.4

3.18 **Bibliography**

SULPHUR.

1. Texas Gulf Company (1953) "Facts about sulphur".

2. Tuller W.N. (1954) "The sulphur data book", pp 96-104. McGraw Hill Book Co. Inc., New York.

3. British Standard Specifications Sulphur BS 4113-1967.

4. Austin G.T.(1986) Shreve's Chemical Process Industries, 5th ed. McGraw Hill Book Co. Inc., New York. Chapter 19: "Sulfur and sulfuric acid". pp 320-345.

5. FPA Information Sheets on Hazardous Materials H14, Sulfur (1973).

6. Chemical Safety Data Sheet SD-74 (1954). Manufacturing Chemists Association, M.C.A., Washington, USA.

SULPHUR OXIDES.

7. Allied Chemical Publication. (1968) "Sulfan, liquid stabilised sulphur trioxide".

8. Bradley G.K. (1982) ICI Mond Division, Technical Dept., P.O. Box 8, The Heath, Runcorn, Cheshire, England. "Treatment of sulfur acids spills". Proceedings of Hazardous Material Spills Conference.

9. Sumner C.A., Pfann J.R. (1973), "Sulfur-trioxide Spill Control" (report on trials in USA in 1973).

10. Stauffer Chemical Co. Westport, Connecticut 06881, USA. Sulfur trioxide product safety information leaflet.

11. Manufacturing Chemists Ass., USA. Chemical Safety Data Sulfur Trioxide, sheet SD-101.

12. The Soap and Detergent Industry Ass., UK (1986). "Recommended safe practices and emergency procedures for sulphur trioxide, oleum and chlorosulphonic acid".

13. Duke J. (1985) (P&G). 78th Annual Meeting of the Air Pollution Control Ass., Detroit, Michigan. "Estimating down wind distances from fuming acid spills".

14. The Safety Practitioner Hazard Data Book. (1980) sheet number 40. "Sulphur oxides and sulphur oxide compounds".

OLEUM 65%, OLEUM 20%, and SULPHURIC ACID.

15. Stanton and Staveley Ltd. (1975). Notes on the properties and handling of oleums (fuming sulphuric acids).

16. Stauffer Chemical Co. Product Safety Information Oleum.

17. Chemical Week, 29 August 1973, report on oleum leak.

18. ICI Ltd. Safety Dept. "65% oleum, codes of practise for chemicals with major hazards".

19. ICI Mond Div. "Sulphuric acid, bulk storage and handling of concentrated sulphuric acid".

20. Darling H.E. (1964) "Conductivity of sulfuric acid solutions". Journal of Chemical and Engineering Data, vol. 9(3), p 421.

21. Record R. (1967). "Conductivity of sulphuric acid". Instrument Engineer, April 1967, p 131.

ORGANIC FEEDSTOCKS.

22. Pardon R.A. (1989) "Manufacturing processes for organic feedstocks", paper presented at Practical Sulphonation Technology Seminar, April 1989 - The Hague, The Netherlands. Center for Professional Advancement (CFPA), PO Box H, East Brunswick, New Jersey 08816, USA.

23. Pardon R.A. (1989) "Organic feedstock quality and specification", paper presented at Practical Sulphonation Seminar, April 1989. See reference 22.

24. Gehrmann M. (1989) "Surfactants on the basis of naturally renewable organic raw materials", paper presented at Practical Sulphonation Seminar, April 1989. See reference 22.

25. Falbe J. (editor) (1986) "Surfactants in consumer products. Theory, technology and application". Springer Verlag, Berlin.
Section 3.1. Raw Materials, pp 24-53.

26. Davidsohn A., Milwidsky B.M. (1978) "Synthetic Detergents". 6th ed. John Wiley & Sons, New York. Raw Materials for Anionic Synthetic Detergents, pp 93-122.

27. Shell Chemical International Trading Company. Shell Centre, London SE1-7PG: High purity linear alkyl benzene derived from SHOP olefins with
the alkyl chain in the range C_{10}-C_{14}.
DOBANE (1983) - Technical Bulletin DI 3.2.1.
DOBANE 102 data sheet UD 103.
DOBANE 103 data sheet UD104.
DOBANE 113 data sheet UD 105.
DOBANE 124 data sheet UD106.
SHOP (Shell Higher Olefins Process) Linear Alpha Olefins, brochure of 1982.
SHOP C_{12}, data sheet UO 101 (1985).
SHOP C_{14} data sheet UO 102.
SHOP C_{16}, data sheet UO 103.
SHOP C_{18}, data sheet UO 104 (1985).
DOBANOL high quality detergent range primary alcohols and primary alcoholethoxylates. Technical Bulletin DI3.2.10 (1983).
DOBANOL 23 (carbon chain 12,13), data sheet DI 2.3.2.
DOBANOL 25 (carbon chain 12,13,14,15), data sheet ZA 133 (1979).
DOBANOL 23-2 (carbon chain 12,13, with a mean 2EO per molecule), data sheet DI 2.4.5 (1979).
DOBANOL 23-3, UD 409, (1983).
DOBANOL 25-3, DI 2.4.8 (1979).

CAUSTIC SODA, SODA ASH and AMMONIA.

28. Staveley Chemicals Ltd. (1975) notes on the properties and handling of caustic soda.

29. ICI-PLC Mond Div. Technical Service Note TS/A/2388 (1984). "Sodium hydroxide. Dissolving caustic soda and handling its solution".

30. ICI-PLC Mond Div. Technical Service Note TS/A (1984) "Sodium hydroxide. Caustic soda liquor".

31.BP Chemicals Ltd. Caustic soda, booklet number JB101/2 (1984).

32. Riedel L. (1950) "Thermal conductivity of aqueous caustic soda solutions". Chem. Ing. Tech. 22 p54.

33. ICI-PLC Mond Div. Technical Service Note TS/A/2152/1 (1982) "Sodium carbonate. Properties, bulk transport, storage & handling".

34. ICI-PLC. Industrial Hazards Bulletin,
Ammonium Hydroxide AG/894.

35. M.C.A., chemical data sheet SD-13 aqueous ammonia.

CHLORINE, HYPO-CHLORITE, HYDROGEN-PEROXIDE.

36. FPA (Fire Protection Ass., 140 Aldersgate Street, London EC1A4HX), information sheets on hazardous materials H39 (1987) Chlorine.

37. UK Chemical Industries Ass. (1984) "Guidelines for bulk handling of chlorine at customers installation".

38.ICI-Mond Div. (1965) Properties and handling liquid chlorine.

39. Allied Chemical Corporation (1961). Liquid Chlorine. Technical and Engineering Service Bulletin no. 7.

40. BP Chemicals, Technigram J103, Sodium Hypochlorite.

41. FPA. Information sheets on hazardous materials H52 (1986). Sodium Hypochlorite.

42. Solvay Bulletin (1960). Chlorine bleach solutions. Bull. 14.

43. FPA. Information sheets on hazardous materials H3 (1987). Hydrogen Peroxide.

44.Interox Chemicals Ltd. (Moorfield road, Widnes, Cheshire, UK). Hydrogen peroxide. P 1.3.1. Manual for handling and storage of 27.5 - 35.0% by weight hydrogen peroxide.

45. Laporte Chemicals Ltd. Hydrogen Peroxide.

46. Schumb W.C. et al. Hydrogen Peroxide. Rheinhold Publishing Company, New York.

ETHANOL.

47. BP Chemicals Ltd. Bulk storage of highly inflammable liquids, ethanol. Technical Booklet SB 116/2.

48. FPA. Information sheets on hazardous materials, ethyl alcohol.

4 PROCESS CHEMISTRY AND PRINCIPLES

4.1 SO_3/air production chemistry

4.1.1 Sulphur combustion with excess dry air to SO_2/"air"

Liquid sulphur is burned to form sulphur dioxide (SO_2) which is subsequently used for the production of SO_3/air. The liquid sulphur is first vaporised and then reacts with oxygen to form SO_2. In practice, vaporisation is accomplished by providing heat by burning part of the sulphur. The sulphur vapour and air are then mixed to allow the oxidation reaction to reach completion.

Table 1 Heat data Sulphur burning

Sulphur ignition temperature °C, appr. 246
Heat of combustion (25 °C):
$S + O_2 \rightarrow SO_2 \quad \Delta H = -9282$ kJ/kg S
Specific heat of sulphur (appr. average over temperature range)

Rhombic S solid	(24.9 - 95.5 °C)	0.75 kJ/kg °C
Monoclinic S solid	(95.5 - 119 °C)	0.75 kJ/kg °C
Liquid	(119 - 140 °C)	0.96 kJ/kg °C
Heat of transformation		
Rhombic → Monoclinic		12.6 kJ/kg
Heat of melting		45 kJ/kg

Specific heat of gases (appr. average over temperature range)

SO_2 gas	0.647 kJ/kg °C
N_2	1.025 kJ/kg °C
O_2	0.916 kJ/kg °C

The approximate actual gas outlet temperature of the sulphur burner can be calculated for various SO_2 volume concentrations in the combustion gas.

The heat of combustion is based on a temperature of 25 °C. Liquid sulphur is fed to the furnace at about 140 °C and therefore the value of $\Delta H = -9282$ kJ/kg has to be corrected for the enthalphy difference between solid sulphur at 25 °C and liquid sulphur at 140 °C:

Solid 1 x 0.75 x 94	70.5
Heat of transformation R → M	12.6
Heat of melting (at 119 °C)	45
Liquid 1 x 0.96 x 21	20.2
	148.3 kJ/kg

i.e. ΔH corrected (9282 - 148) = -9134 kJ/kg.

The exhaust gas temperature can now be calculated for various vol. % SO_2 in the exhaust gas. Assume for example a SO_2 vol.% of 10%.

Reaction equation:
$S + (O_2)_a + N_2 \rightarrow SO_2 + (O_2)_b + N_2$

(The air is bone dry and therefore traces moisture can be neglected.)

1 kg S corresponds to
1/32 kmole SO_2 or 10/32 kmol
combustion gas assuming 10 vol. % SO_2.

Mass balance in kmole:
$N_2 + (O_2)_b = 0.9 \times 10/32 = 9/32 = 0.281$

$N_2 = 4\ (O_2)_a$

$(O_2)_a - (O_2)_b = 1/32 = 0.031$

From above equations:

$(O_2)_a = 0.0624$ kmole
$N_2 = 0.2496$ kmole
$(O_2)_b = 0.0314$ kmole

in mass units:
1 kg S + 2 kg O_2 + 7 kg N_2 $\rightarrow$ 2 kg SO_2 + 1 kg O_2 + 7 kg N_2

<u>Heat balance (base temperature 140°C)</u>

$\Delta H = -9134$ kJ/kgS

Heat in:
$2 \times 0.916 \times (25-140) + 7.00 \times 1.025 \times (25-140) + 9134$ is

Heat out:
$2 \times 0.647(T-140) + 1 \times 0.916(T-140)$
$+ 7 \times 1.025(T-140)$

Heat in = Heat out $\rightarrow$ T = <u>1003°C</u>

The same balances can be made for different SO_2 concentrations, e.g. for 3, 5, 7% SO_2 in exhaust gas. Results are summarised in table 2.

Table 2 Approximate burner gas outlet temperatures related to SO_2 concentrations (vol.%) in exhaust gases

vol % SO_2	burner gas outlet Temp (°C)
10	1003
7	735
5	524
3	323

N.B. In the furnace itself local spots within the burner may be considerably hotter.

The practical values of sulphur burner outlet temperatures will be marginally lower because of heat losses from the lagged furnace to the ambient atmosphere. The general rule of thumb is that the outlet temperature corresponds to the vol% SO_2 in the gas x 100 e.g. 7% SO_2 gas ex furnace corresponds to a burner outlet temperature of about 700°C. However, it should be remembered that when the process air is preheated (MM, Mazzoni designs) then the exhaust gas temperature will be roughly 10% higher.

The heat produced by the combustion of sulphur can be utilized in boilers to generate steam for melting the sulphur or for steam supply purposes around the plant; or to heat ambient air required for regenerating the silica gel beds used for drying process air to a dew-point of about - 60°C.

4.1.2 Conversion of SO_2 to SO_3

The reaction from SO_2 to SO_3 is an exothermic reversible reaction:

$$SO_2 + \tfrac{1}{2}O_2 \xrightleftharpoons{V_2O_5 \text{ catalyst}} SO_3 + \text{heat}$$

$\Delta H = -3043$ kJ/kgS at 427 °C

Both thermodynamic and stoichiometric considerations are involved in maximising the formation of SO_3. To decide how to optimise the equilibrium, consider the van 't Hoff-Le Chatelier Principle, which states that when an equilibrium system is subjected to stress, the system will tend to adjust itself in such a way to partly relieve the stress. The stresses are in the form of an increase or decrease in the temperature, pressure or concentration of a product or reactant. In all catalytic reactions, the function of the catalyst is to increase the rate of the reaction.

The reaction rate increases as the temperature rises but then begins to slow down as equilibrium is approached. The reaction essentially stops when about 60 to 70% of the SO_2 has been converted, at a temperature in the vicinity of 600°C. Then the gas, before it enters the second fixed catalyst bed, is cooled in a heat exchanger or fresh dry air is added to reduce the temperature, to about 430°C. The yields using this procedure for a 4 bed absorption converter without air dilution are 97 to 98%.

Increasing the system pressure will force the system to form more product SO_3, since it occupies less volume and hence will result in a lower pressure (van 't Hoff - Le Chatelier Principle). Super-atmospheric plants are not suitable for sulphonation applications.

Based on the fundamental chemical principles it is possible to achieve high overall conversions by using a multipass converter wherein, at an entering temperature of 425 - 440 °C, the major part of the conversion (60 - 75%) is obtained in the first catalyst bed with an exit temperature of 600 °C or more. The successive lowering of the temperature between stages ensures an overall higher conversion. Conversion in a 4-stage converter of a SO_3 production plant is 98 %. Table 4 gives the temperature and percent conversion for each stage of a 4-pass converter.

Two grades of V_2O_5 catalyst are charged into the converter, a less active but harder type being used in the first bed of the converter and a more active but softer type in subsequent passes. These catalysts have a long lifetime (up to 20 years) and are not subject to poisoning, except by Arsenic, Te, Se and fluorine which damages the siliceous carrier. Dust must be removed from the catalyst occasionally by screening. Vanadium catalyst or dust must not be inhaled, ingested or allowed to contact cuts or open sores. Instructions for handling and sieving V_2O_5 catalyst are given in appendix 3.

Fig. 6 Equilibrium data for gas from sulphur burning

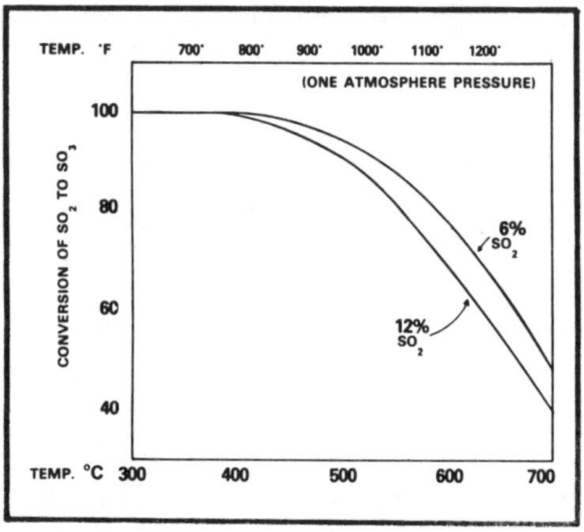

Table 3 Equilibrium constants for SO_2 oxidation at 1 atm. total pressure

Temp [°C]	K_p *
400	397
500	48.1
600	9.53
700	2.63

$*K_p = P(SO_3)/\{P(SO_2)xP(O_2)^{½}\}$ (pressures in atm)

Table 4 Temperatures and conversions in each stage at a Monsanto converter, using relatively rich SO_2 gas from sulphur (8 vol % SO_2)

Location		Temperature °C	Equivalent conversion%
Gas entering first pass		410.0	74
Gas leaving first pass		601.8	
Rise in temperature	191.8		
Gas entering second pass		438.0	18.4
Gas leaving second pass		485.3	
Rise in temperature	47.3		
Gas entering third pass		432.0	4.3
Gas leaving third pass		443.0	
Rise in temperature	11.0		
Gas entering fourth pass		427.0	1.3
Gas leaving fourth pass		430.3	
Rise in temperature	3.3		
Total rise in temperature	253.4		
Total conversion			98.0

4.2 **Sulphonation chemistry**

4.2.1 Alkylbenzene sulphonation

The main sulphonation reactions can be written as follows:

$RC_6H_5 + 2\ SO_3 \rightarrow RC_6H_4SO_2OSO_3H$ (fast)
pyrosulphonic acid

$RC_6H_4SO_2OSO_3H + RC_6H_5 \rightarrow 2\ RC_6H_4SO_3H$ (slow)

The reaction is highly exothermic with a ΔH of -170 kJ/gmol.
Due to the slow second reaction ageing is required to convert the formed pyrosulphonic acid with alkylate to the desired end product LABSA.

There are two main side reactions:
anhydride formation and sulphone formation

Sulphonic anhydride

$$RC_6H_4SO_2OSO_3H + RC_6H_4SO_3H \rightarrow RC_6H_4SO_2OSO_2C_6H_4R + H_2SO_4$$

In the presence of excess alkylbenzene (ageing):

$$RC_6H_4SO_2OSO_2C_6H_4R + H_2SO_4 + RC_6H_5 \rightarrow 3\ RC_6H_4SO_3H$$

In the presence of water (stabilisation):

$$RC_6H_4SO_2OSO_2C_6H_4R + H_2O \xrightarrow{H^+} 2\ RC_6H_4SO_3H$$

Sulphonic anhydride formation is promoted by:
- high excess SO_3 (mole ratio);
- rapid SO_3 addition (short residence time falling-film reactors);
- feedstock type (linear alkylates).

The formed anhydride can be transformed to the desired LABSA during ageing and notably during the hydrolysis step (stabilisation).

Sulphone

$$RC_6H_4SO_2OSO_3H + RC_6H_5 \rightarrow (RC_6H_4)_2SO_2 + H_2SO_4$$

Sulphones are a constituent of free oil at a level of appr. 25 to 33% and cannot be eliminated by ageing or any other step i.e. sulphone competes directly with the desired main reaction.

Sulphone formation is promoted by the following conditions:
- high temperatures, because the sulphone formation reaction has a higher activation energy than the main sulphonation reactions;
- probably promoted by a high concentration of H^+ ions.

Anhydride and sulphone formation are not the only side- reactions. Other side-reactions lead to coloured products (excess SO_3, high temperatures), traces of disulphonic acid, olefins and oxidation products.

The sulphuric acid concentration should be at that required for the particular product application.

Sources of sulphuric acid in LABSA:

(i) Moisture in feedstock

0.1% H_2O $\rightarrow$ 0.5% H_2SO_4 (appr.)

(ii) Insufficient ageing

Sulphonic anhydride is reacting with H_2SO_4 and LAB, forming LABSA (see before). Insufficient ageing will leave this reaction incomplete.

(iii) <u>Premature stabilisation (hydrolysis)</u>

Not all pyrosulphonic acid may be converted in LABSA (slow reaction). Then

$$RC_6H_4SO_2OSO_3H + H_2O \rightarrow RC_6H_4SO_3H + H_2SO_4$$

(iv) <u>Sulphone formation</u>

See above.

(v) <u>Colour formation</u>

Hydrogen abstraction ultimately gives H_2SO_4.

<u>NOTE</u>:
Some H_2SO_4 can be removed by long ageing periods (hours) providing that some free LAB is present.

From the above reaction chemistry the following sulphonation conditions can be understood:

- correct mole ratio SO_3 : LAB (1.01-1.03 : 1);
- SO_3-in-air concentration (5.0-7.0 vol %)

<u>NOTE</u>:
Low SO_3 levels in air result in better colour, lower H_2SO_4, lower sulphones but possibly longer ageing times may be required;

- good cooling to keep the temperature in the 45-50°C region.

* <u>Ageing</u>

- Approximately 30 minutes (dependent on sulphonation conditions).
- Plug flow characteristics of ageing reactor.

* <u>Stabilisation (hydrolysis)</u>

- Avoid excess water.
- Add sufficient water (normally 1% w/w) to transform anhydrides to avoid pH drift of pastes (check by Karl Fischer titration).

Reaction scheme of AB sulphonation

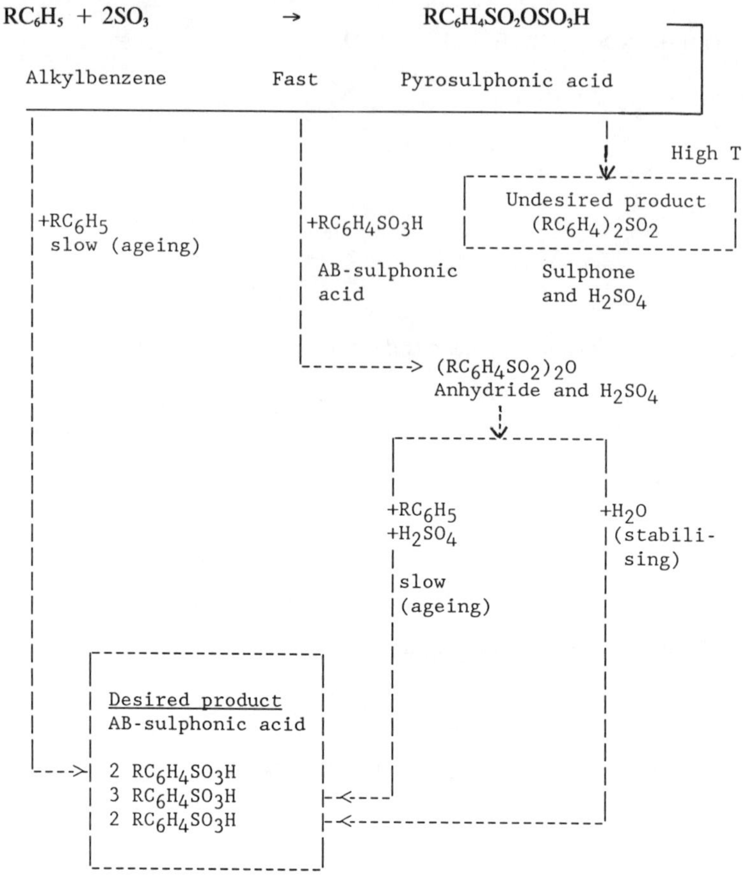

4.2.2 Primary alcohol sulphation

The mechanism for alcohol sulphation is thought to be similar to that for LAB sulphonation, involving the rapid formation of a metastable pyrosulphate species which decomposes less fast but still rapidly to alkyl hydrogen sulphate ("acid sulphate"):

$$ROH + 2\,SO_3 \xrightarrow{\text{fast}} ROSO_2OSO_3H$$

alkyl hydrogen pyrosulphate

less fast

$ROSO_2OSO_3H$ + ROH → 2 $ROSO_3H$
alkyl hydrogen sulphate

The reaction is highly exothermic with a ΔH of -150 kJ/gmole.

An important feature of sulphation chemistry is the thermal instability of the "acid sulphate", which breaks down to a mixture of products including the parent alcohol, the dialkyl sulphate ($ROSO_2OR$), the dialkyl ether (ROR), isomeric alcohols, olefins ($R'CH=CH_2$) and esters ($ROSO_3R$). Because of the thermal instability of the "acid sulphate" it is necessary to avoid high sulphation temperatures and to neutralise the acid sulphation product soon after its formation. An ageing time of about 1 minute at 30-50°C is adequate for the second reaction where the desired alkyl hydrogen sulphate is formed. In practice the minimum sulphation temperature feasible is determined by the need for the feedstock and reaction mixture to be mobile liquids.

From the above reaction chemistry the following sulphation conditions can be understood for alcohol sulphation:

- mole ratio: SO_3 : alcohol 1.02-1.03 : 1;
- 3-5% SO_3 mole fraction in process air;
- organic feedstock entrance temperature 30°C for C_{12} range (slightly higher than melting temperature);
- sulphonation temperature appr. 40°C (35-40°C for C_{12}-C_{14} alcohols) and appr. 50°C (45-55°C) for C_{16}-C_{18} alcohols;
- neutralisation of organic acid within 1 minute after sulphonation reactor. A separate ageing step is not needed under commercial reactor conditions;
- short residence time falling-film reactor required to avoid thermal breakdown of $ROSO_3H$.

Reaction scheme of primary alcohol (PA) sulphation

```
                       very fast
ROH + 2SO3 ----------> ROSO2OSO3H
                              Alkyl hydrogen pyrosulphate
                              ---------------------------
                              | Desired product         |
                     fast     |-------------------------|
ROSO2OSO3H + ROH  <=====>     | 2ROSO3H                 |
                     slow     | Alkyl hydrogen sulphate |
                              ---------------------------
                                           |
 ---------------- T and t ------------------
 |
 |          ------------------------------------------
 |          |          UNDESIRED BY-PRODUCTS          |
 |          |-----------------------------------------|
 |------->  | Parent alcohol, but secondary alcohols  |
 |          | can also be observed, usually pre-      |
 |          | dominant by-product levels below 2%     |
 |          | can be obtained                         |
 |          |                                         |
 |------->  | ROSO2OR, di-alkyl sulphate, traces      |
 |          |                                         |
 |------->  | ROR, di-alkyl ethers, low levels        |
 |          | appr. 0.2%                              |
 |          |                                         |
 |------->  | Olefins, R'CH=CH2, usually below 0.1%   |
 |          |                                         |
 -------->  | Esters ROSO3R, below 0.5%, often        |
            | below 0.1%                              |
            |                                         |
            ------------------------------------------
```

4.2.3 Ethoxylated alcohol sulphation

Theoretically, the mechanism for ethoxylated alcohol sulphation is similar to primary alcohol sulphation, involving the rapid formation of a metastable product. The stoichiometry of this almost instantaneous and highly exothermic initial reaction corresponds again to more than one molecule of SO_3 per molecule of feedstock:

$$\text{R-O-CH}_2\text{CH}_2(\text{OCH}_2\text{CH}_2)_n\text{OH} + 2\ \text{SO}_3 \xrightarrow{\text{fast}} \text{RO}^+(\text{SO}_3^-)\text{CH}_2\text{CH}_2(\text{OCH}_2\text{CH}_2)_n\text{OSO}_3$$

$$\text{R-O}^+(\text{SO}_3^-)\text{-CH}_2\text{CH}_2(\text{OCH}_2\text{CH}_2)_n\text{OSO}_3\text{H} + \text{R(O-CH}_2\text{CH}_2)_{n+1}\text{-OH} \xrightarrow{\text{less fast}} 2\ \text{R(OCH}_2\text{CH}_2)_{n+1}\text{OSO}_3\text{H}$$

$\Delta H = -150$ kJ/gmole

The desired ethoxylate acid sulphate product formed is less prone to thermal decomposition than the primary alcohol sulphation acid product. This is usually attributed to the moderating influence of the ethoxylate chain on the reactivity of SO_3, together with stability of the ethoxylate acid sulphate product. The alcohol ethoxylate acid sulphate requires rapid and efficient neutralisation to maintain good product quality (preventing by-products, notably dioxane) and colour.

An important side reaction is dioxane formation:

$$R(OCH_2CH_2)_nOSO_3H \xrightarrow{H^+} R(OCH_2CH_2)_{n-2}OSO_3H + \underset{\text{1,4-dioxane}}{(CH_2CH_2O)_2}$$

1,4-dioxane has a cyclic structure:

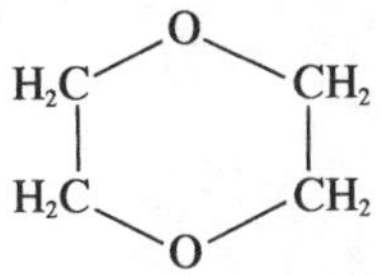

Dioxane forms by the chemical cleavage of two molecules of ethylene oxide from the parent ethoxylated alcohol. Dioxane is the undesirable by-product. The amount of dioxane ranges from traces to hundreds, even thousands, ppm (mg/kg) depending on raw material quality and sulphonation/neutralisation process conditions.

Dioxane formation is favoured by:

1. Excess of SO_3 (poor mole ratio control) i.e. high conversion, low free oil.

2. High temperatures, when the exothermic reaction is poorly controlled.

3. Long (minutes) organic acid ageing times (direct function of Temp.).

4. Moisture in feedstock causing the presence of strong acids, i.e. sulphuric acid.

5. Branching of the alkyl chain OXO alcohols.

6. Longer EO chains in the ethoxylated alcohol feedstock i.e. 3-EO's present higher dioxane levels than 2-EO's, all other parameters assumed identical.

Dioxane will exist in the protonated form or as an SO_3 complex in the acid mix from which it will be liberated by alkali. The following process conditions have to be established for good quality ethoxylated alcohols of the type C_{12}-$C_{14}/_{15}$ 2-3 EO.

1. Mole ratio 1.01-1.02 : 1, SO_3 : ethoxylate. Higher ratios would lead to higher levels of dioxane (> 100ppm). Lower mole ratios would affect conversion (free oil > 3% on 100% AD).

2. Moderate effectively the temperature rise due to the exothermic reaction to control dioxane formation and colour:

   (i) SO_3 in air, 3% vol. max.

   (ii) organic feedstock temperature: 30°C

   (iii) throughput: 85% weight percentage of nominal for alkylbenzene (this corresponds with about 50-70% on a mole basis → less heat of reaction → lower temperature peak)

   (iv) optimal cooling water temperatures and cooling water flows in upper and lower reactor cooling circuits.

Reaction scheme of ethoxy alcohol sulphation

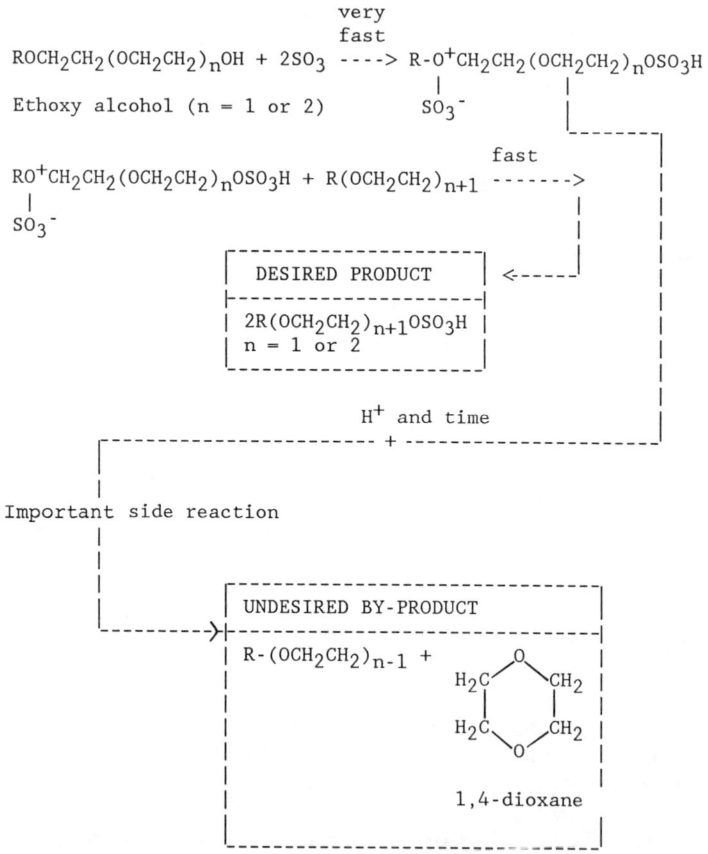

4.2.4 Alpha-olefin sulphonation (AOS)

Only the alpha-olefins are sulphonated commercially to make alpha-olefin sulphonate (AOS). The chemistry of alpha-olefin sulphonation is usually described in terms of 3 stages.

(i) The initial sulphonation reaction involves the formation of B-sultones. This initial step is so fast as to be almost instant. Reaction of the initially formed B-sultone with more SO_3 is competitive with sulphonation of the olefin. This side reaction produces a by-product believed to be a cyclic pyrosulphonate ester i.e. a pyrosultone:

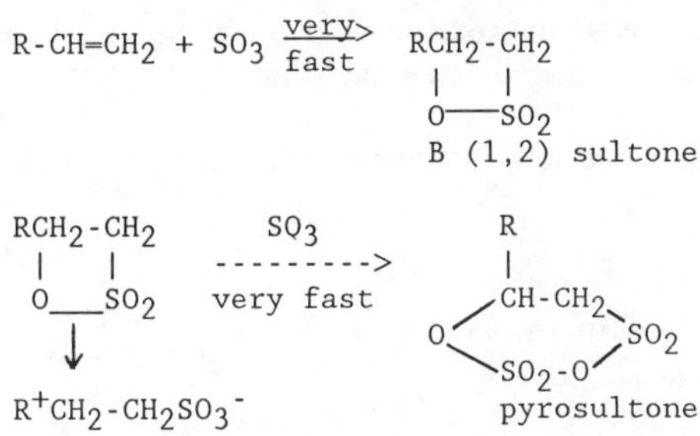

These reactions are highly exothermic with $\Delta H = -210$ kJ/gmole

(ii) In the case of alpha-olefin sulphonation, the pyrosultone is metastable, decomposing on ageing to alkene sulphonic acids and releasing its extra SO_3 which can then react with any olefin remaining. On ageing the B-sultone isomerises to a mixture of alkene sulphonic acid and C- and D-sultones. If this ageing is not carried out, the neutralised product contains a high proportion of 2-hydroxysulphonate, which is undesirable due to its poor solubility:

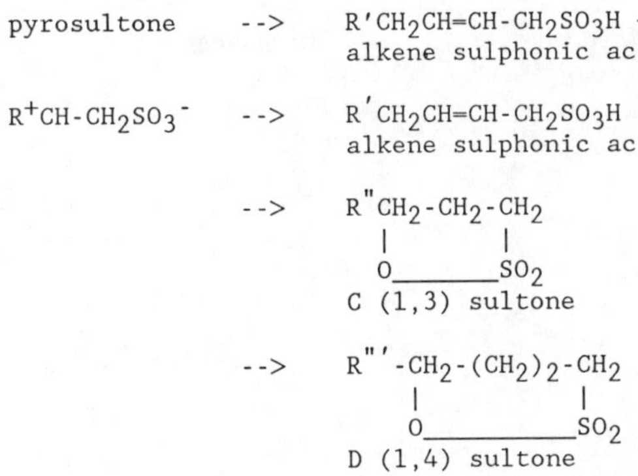

After a residence time of 15-20 minutes at slightly elevated temperature (35°C), the B-sultones have virtually disappeared and the D-sultones are only present at low levels. At this point where neutralisation followed by hydrolysis (see 3.3.4) is usually carried out, the product composition can be depicted as follows:

-	Alkene sulphonic acid	appr. 30%
-	C(1,3) alkane sultone including small amounts of D(1,4) alkane sultone	appr. 50%
-	Di-sultone	appr. 10%
-	Disulphonic acids)	appr. 10%
-	Sultone sulphonic acids)	

(iii) Side reactions:
Several side-reactions begin to predominate over sultone formation as the reactions proceed. These reactions basically take the form:

- isomerisation of the sultone intermediate to form the desired alkene sulphonic acid;

- disulphonic acid formation

$$\begin{array}{l} R\text{-}CH{=}CH\text{-}CH_2\text{-}(CH_2)_xSO_3H \\ \qquad\quad | \\ \qquad\ SO_3H \end{array}$$

- oxidation/reduction reactions which lead to formation of SO_2 in the exit gas.

Because of disulphonation and oxidation reactions, an excess of SO_3 must be used.

Operating conditions for alpha-olefin sulphonation:

-	mole ratio SO_3/olefin	1.06-1.08
-	SO_3 concentration	2.5 -4.0 vol %
-	cooling water chilled	15°C
-	digestion time before neutralisation/hydrolysis	5-10 minutes
-	digestion temperature	30-35°C

Reaction scheme of alpha-olefin sulphonation

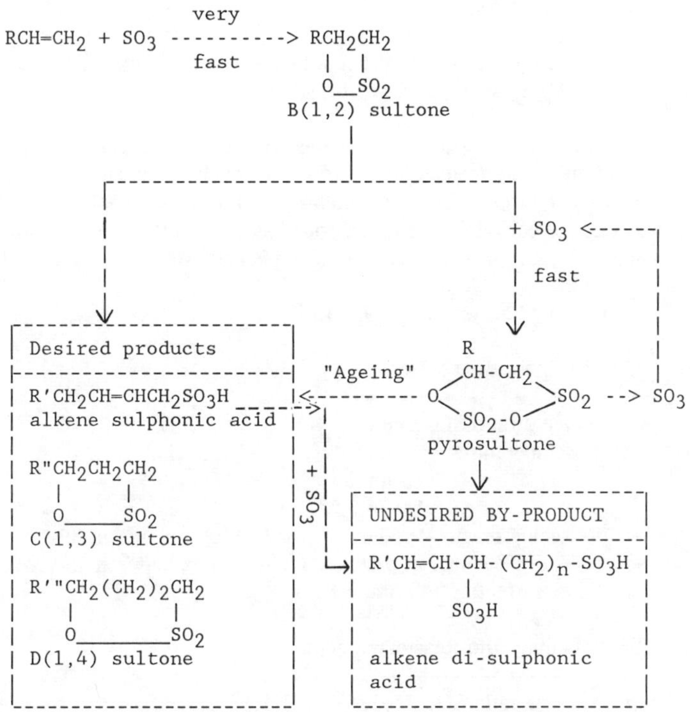

4.2.5 Fatty acid methyl ester (FAME) sulphonation

The chemistry of FAME sulphonation is complicated and not yet fully elucidated but may be summarised as depicted in the following reaction scheme.

A. The initial reaction between FAME and SO_3, although fast, is far from instantaneous. Two intermediate products are formed:

I. probably $RCHCO\text{-}O\text{-}SO_2OMe$ (2:1 stoichiometry), with SO_3H on the CH

II. a mixture of intermediates with an overall stoichiometry 3:1

Thus, if a mole ratio of about 1.2:1 of SO_3 to ester is used, the initial reaction gives about 50% conversion of starting material, the products being intermediates I and II. On ageing, these intermediates release SO_3, which reacts with the remaining ester, and the final product is a mixture of FAMES acid (FAMES = Fatty Acid Methyl Ester Sulphonate) and SFAS acid (SFAS = Sulphonated Fatty Acid Sodium salt, RCH-COONa, with SO_3Na on the CH).

On neutralisation an active is obtained with appr. the mole ratio composition 80:20 FAMES:SFAS. Sodium methyl sulphate (CH_3OSO_3Na) is also produced in a molar quantity roughly equivalent to the SFAS.

Treatment of the acid mix with excess methanol before neutralisation leads to an active which is almost completely FAMES, with very low SFAS levels.

Me-ester sulphonation has to be carried out at relatively high temperatures as the initial reactions and the decomposition of intermediate products are relatively slow compared with sulphonation reaction rates for alkyl benzenes, primary alcohols, ethoxylated alcohols and alpha-olefins. The required ageing time for conversion of the intermediates to FAME Sulphonation Acid is long (about 45 minutes at 85°C). It is not possible to sulphonate Me-esters without an excess of SO_3.

Operating conditions for Me-ester sulphonation:

(i) CSTR sulphonation system

- mole ratio SO_3: Me-ester — 1.25-1.3:1.0
- temperatures reactor 1 — 55°C
  reactor 2 — 65°C
  reactor 3 — 80°C
- ageing reactor 4, 5, 6 — 60 minutes at 80°C
- with bleaching in paste (NaOCl) colours between 150-250 klett are possible (5% solution, 40 mm cell).

(ii) FFR (Falling-film Reactor) system

- mole ratio SO_3: Me-ester 1.2:1
- reaction temperature in short residence time reactor appr. 90°C, cooling water about 80-85°C in lower part and 40-50°C in upper part
- ageing reactor post-FFR required residence time 0.5 h at 90-95°C, plug flow conditions
- with double bleaching step (H_2O_2 in acid, NaOCl in paste) colours of 100-150 klett are possible.

Reaction scheme of FA-Me-ester sulphonation

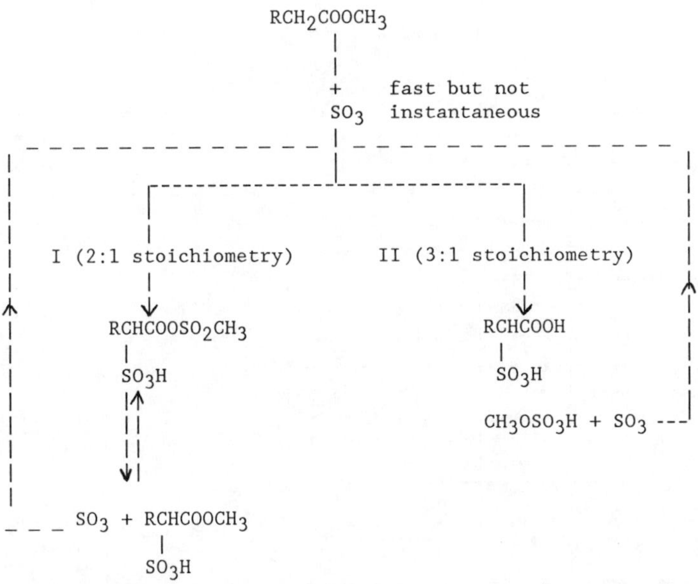

4.2.6 Summary sulphonation reaction process chemistry

(i) Sulphonation of the common feedstocks proceeds with a highly exothermic instantaneous initial reaction, followed by a fast but not instant step, also highly exothermic. The second reaction does not always proceed to completion (e.g. LAB, FAME) in the lower zone of a short residence time falling-film reactor (FFR). For these organic feedstocks ageing under well-defined conditions of temperature and reaction time is required.

(ii) Parallel and side reactions occur resulting in undesired by-products including dark coloured organic compounds take place, promoted by high temperatures and long reaction times.

(iii) The viscosity of the organic increases sharply particularly from 70-100% conversion.

The combination of highly exothermic reactions with a sharp increase in viscosity as conversion proceeds controls reactor design and operational conditions in full-scale operations. The art of sulphonation is to maintain the optimal reaction temperature and reaction time, resulting in products with small amounts of by-products and good colours.

Table 5 Summary of process chemistry parameters for sulphonation of various organic feedstocks

FEEDSTOCK	ABBR.	MW	MOLE RATIO SO_3/ORG.	$-\Delta H$ kJ/gmole VOL/VOL SO_3/AIR	REACTION T [°C]	AGEING T [°C] t [min]	REMARKS
Linear Alkyl benzene-L Linear Alkyl benzene-H Branched Alkyl benzene	LAB LAB BAB	230-240 245-260 245-260	) 1.01-1.03) SO_3) 1.0 org.	170 5-7%	45-50°C	45-50°C 30 min.	After ageing stabilisation to convert anhydride appr. 1% H_2O, LAB-SA, stable product
Lauryl C_{12}-C_{14} alc. Tallow C_{16}-C_{18} alc.	LA TA	200-220 250-265	1.02-1.03 SO_3 1.0 org.	150 3-5%	35-40°C 60-65°C	-- --	Acid product to be neutralised directly after sulphonation
Lauryl ethers 2EO Lauryl ethers 3EO	LE-2 LE-3	290-310 330-350	1.01-1.02 SO_3 1.0 org.	150 3% max.	40°C	--	Acid product to be neutralised directly after sulphonation
Alpha olefins 1-dodecene Alpha olefins 1-hexadecene	AO-C_{12} AO-C_{16}	170 225	1.06-1.08 SO_3 1.0 org.	210 2.5-4.0%	30-35°C	30-35°C 25 min.	Acid product to be neutralised/hydrolised after sulphonation
Fatty Acid Me-ester	FAME-C_{18}	300	1.20 SO_3 1.0 org.	5-7%	80-85°C	90-95°C 30 min.	Bleaching of acid and neutralised product required

4.3 Neutralisation chemistry

4.3.1 Introduction

The process consists of an exothermic reaction between a neutralising agent and either sulphonic acid (ex LAB, alpha-olefins, FAME) or acid sulphate (ex primary alcohols, ethoxylated alcohols). Neutralisation can be carried out after prolonged storage, if the acid stability permits (LABSA, FAME-SA).

The most commonly used neutralising agents are caustic soda (NaOH) and sodium carbonate (soda ash, Na_2CO_3) with less common ones being ammonia solution (NH_3), mono- and diethanolamines. Due to the strong base character of NaOH, the heat of neutralisation, in the order of 100 kJ/gmole, is higher than that obtained with the other agents.

Efficient mixing of the organic acid and the alkaline phase is essential to promote the instantaneous reaction and to avoid undesired degradation reactions in isolated acid spots and pH drift occurring in the neutralised product storage tanks, arising from small amounts of unneutralised acid lumps. This mixing operation requires high-shear mixers as the formed reaction product, the so-called paste, is highly viscous at low shear rates.

Small temperature variations ($+5^{\circ}$C) can change the phase structure of the paste dramatically, resulting in undesired rheological structures like gels. Good temperature control is therefore key in neutralisation reactions.

pH control is also essential for good quality products. Too low a pH value (<6) can result in product decomposition (hydrolysis of alcohol sulphates, ether sulphates and FAMES) leading to a further pH drop in the storage tanks. Too high pH values (>11) leave too much caustic soda in the paste, with negative consequences in following processing stages (e.g. NSD powder properties). pH control within the range of 7-9 is impossible for strong acid/strong base systems e.g. LABSA with NaOH. However, it is possible to control the pH in the 9-10 region where the titration curve becomes less steep.

4.3.2 Alkylbenzene sulphonic acid (LABSA) neutralisation

The neutralisation reaction with diluted caustic can be written as follows:

$$RC_6H_4SO_3H + NaOH \rightarrow RC_6H_4SO_3Na + H_2O$$

LABSA LABS ("AD")

The reaction is highly exothermic due to the heat of neutralisation and the heat of dilution of strong acids and a strong base (50% caustic is the currently available strength). At present there is little theoretical data on the enthalpies involved in the neutralisation reaction between sulphonic acid and sodium hydroxide solution. Values of about 100 kJ/gmole have been found experimentally. The aim of the following exercise is to specify the individual heats involved and calculate ΔH_{total} for various caustic strengths and compare these various ΔH_{total} with the practical value of 100 kJ/gmole.

The following reactions and heats are involved:
ΔH_1: heat of dilution of sulphonic acid in water (to infinite solution)
ΔH_2: heat of neutralisation of LABSA
ΔH_3: heat of dilution of 100% H_2SO_4 to infinite dilution
ΔH_4: heat of neutralisation of H_2SO_4
ΔH_5: heat of dilution of 50% NaOH to infinite dilution

In table 6 and table 7 heats of neutralisation and dilution are summarised respectively.

Table 6 Heats of neutralisation (excluding dilution/solution effects)

Acid	Base	Heat evolved kJ/g mole acid
Sulphonic	Caustic	61.4
Sulphuric	Caustic	112.0
Sulphonic	Soda ash	12.6
Sulphuric	Soda ash	25.2

Table 7 Heats of dilution (to infinite dilution)

Acid	Base	Heat evolved kJ/g mole	Heat evolved kJ/kg H_2SO_4 NaOH
Sulphonic		39.3	
98% H_2SO_4		96.9	988.5
	Solid NaOH	42.3	1057.5
	50% NaOH		488.4
	40% NaOH		265.6
	30% NaOH		86.8
	20% NaOH		-3.5

Working on the basis of 1.0 kg of "Dobane 113" sulphonic acid (MW 320) having the following composition:

	%	Mass (g)	g moles (n)
Dobane 113 SA	96	960	3
H_2SO_4	2	20	0.2
free oil	2	20	
	100	1000	

The acid will be neutralised with caustic of 50%, 30% and 20% to a paste of 45% AD strength. The overall equations for the 3 cases are as follows:

Starting with 50% caustic (bracketed figures indicate H_2O present above stoichiometric calculations).

$$RSO_3H + H_2SO_4 + NaOH + (H_2O) \rightarrow RSO_3Na + Na_2SO_4 + H_20$$

	RSO_3H	H_2SO_4	NaOH	(H_2O)	RSO_3Na	Na_2SO_4	H_2O
Mass	960	20	136		1023	28.4	61.2
MW	320	98	40		341	142	18
N	3	0.2	3.4		3	0.2	3.4
H_2O in caustic				(136)			(136)
AD dil'n. H_2O				(1008)			(1008)

$$RSO_3H + H_2SO_4 + NaOH + (H_2O) \rightarrow RSO_3Na + Na_2SO_4 + H_20$$

30% caustic

	RSO_3H	H_2SO_4	NaOH	(H_2O)	RSO_3Na	Na_2SO_4	H_20
Mass	960	20	136		1026	29	61
H_2O in caustic				(317)			(317)
AD dil'n. H_2O				(827)			(827)

20% caustic

	RSO_3H	H_2SO_4	NaOH	(H_2O)	RSO_3Na	Na_2SO_4	H_20
Mass	960	20	136		1026	29	61
H_2O in caustic				(544)			(544)
AD dil'n. H_2O				(600)			(600)

The calculation for ΔH 1-5 can be made for the 3 cases: neutralisation with 50% caustic, 30% caustic and 20% caustic.

		ΔH value kJ	
ΔH_1 (heat of dilution LABSA)			
39.3 kJ/g mole x 3 g mole		117.9	
ΔH_2 (heat of neutralisation LABSA)			
61.4 kJ/g mole x 3 g mole		184.2	
ΔH_3 (heat of dilution H_2SO_4			
988.5 kJ/kg x 0.02 kg		19.8	
ΔH_4 (heat of neutralisation H_2SO_4)			
112 kJ/g mole x 0.2 g mole		22.4	
	Sub total	344.3	
ΔH_5 (heat of dilution caustic)			
50% caustic			
488.4 kJ/kg x 0.136 kg			66.4
	ΔH total		410.7
30% caustic			
86.8 kJ/kg x 0.136 kg			11.8
	ΔH total		356.1
20% caustic			
-3.5 kJ/kg x 0.136 kg =			negligible
	ΔH total		344.3

So depending on the strength of caustic to be used, the total ΔH value becomes:

ΔH total 50% caustic 410.7 kJ → 128.3 kJ/g mole acid mixture
ΔH total 30% caustic 356.1 kJ → 111.3 kJ/g mole acid mixture
ΔH total 20% caustic 344.3 kJ → 107.6 kJ/g mole acid mixture

The calculated figures are somewhat higher than the experimental values, probably due to the fact that ΔH_1, heat of dilution of LABSA to infinite dilution, is used. In practice the dilution is not infinite so that ΔH_1 has to be multiplied by a factor between 0.7 and 0.9.

For heat balances required to design neutralisation heat exchangers and operate neutralisation plants, the calculated figures will be conservative and safe.

The heat effects of the reaction between a Na_2CO_3 solution and sulphonic acid are at least a factor 4 smaller than the reaction with a caustic solution:

$$2RC_6H_4SO_3H + Na_2CO_3 \rightarrow 2RC_6H_4SO_3Na + H_2O + CO_2$$

The resulting AD paste for LABSA has 1-1.5% Na_2SO_4 in the paste, resulting from 0.5-1.0% H_2SO_4 in LABSA, reducing the viscosity of the neutralised product. Depending on sulphonic acid quality and Na_2SO_4 level, pastes in the range of 45-60% AD can be produced, transported and stored.

4.3.3 Alkyl hydrogen sulphate ($ROSO_3H$) and alkyl ether hydrogen sulphate $R(OCH_2CH_2)_nOSO_3H$ neutralisation

Neutralisation agents may be caustic soda (predominantly), ammonia, monoethanolamine and triethanolamine. Examples of reaction equations:

$$ROSO_3H + NaOH \rightarrow ROSO_3Na + H_2O$$

$$R(OCH_2CH_2)_nOSO_3H + NH_4OH \rightarrow R(OCH_2CH_2)_nOSO_3NH_4 + H_2O$$

The reaction with diluted caustic to form a paste of 27-30% AD from $ROSO_3H$ and 27-30% or 70% AD from $R(OCH_2CH_2)_nOSO_3H$ is again highly exothermic. Taking into account acid dilution and neutralisation effects, values of 115-125 kJ/gmole have been measured.

Too low pH either overall or in acid spots in the event of poor micro-homogenisation of acid in alkaline phase, should be avoided to prevent hydrolysis of the organic acid product to alcohol (alcohol ether) and sulphuric acid, which will catalyse further hydrolysis:

1. $R\text{-}OSO_3H + NaOH \rightarrow ROSO_3Na + H_2O$
   desired reaction

2. $R\text{-}OSO_3H + H_2O \rightarrow ROH + H_2SO_4$
   hydrolysis in local absence of caustic

3. $2ROSO_3Na + H_2SO_4 \rightarrow Na_2SO_4 + 2RO\text{-}SO_3H$
   sulphuric acid is catalysing hydrolysis reaction

4. $ROSO_3H + H_2O \rightarrow$ etc.

Initial hydrolysis would therefore lead to further hydrolysis and pH drop in storage tanks, resulting in a product that is difficult to recover and may cause irreparable damage (corrosion) to tanks, pipelines and pumps. Therefore the pH must be kept high (9-11) to avoid acid material entering bulk storage (steep titration curve). If for product formulation requirements a product of pH 6-7 is essential, the use of buffers, e.g. phosphoric acid or citric acid, is recommended.

4.3.4 Neutralisation/hydrolysis of alkene sulphonic acid and sultones

Neutralisation of the sulphonation product from alpha-olefins is more complex than neutralisation of the corresponding products of alkylbenzenes. This is because the SO_3-alpha-olefin acid product contains about 50% free sulphonic acid, the rest being C(1,3)- and D(1,4) sultones, assuming that with acid ageing the B(1,2) sultones have disappeared. In the case of alpha-olefins an excess of caustic (1.5-2.0% excess) must be added to neutralise both the sulphonic acid initially present and that formed on subsequent hydrolysis of the C(1,3)- and D(1,4) sultones. The sultones (ring structured esters) cannot be converted into their proper salts by a simple neutralisation but need a hydrolysis step.

Commonly employed base hydrolysis of sultones at 150°C leads to about 60-70% hydroxyalkane sulphonate and 30-40% alkane sulphonate.

Chemical reaction equations:

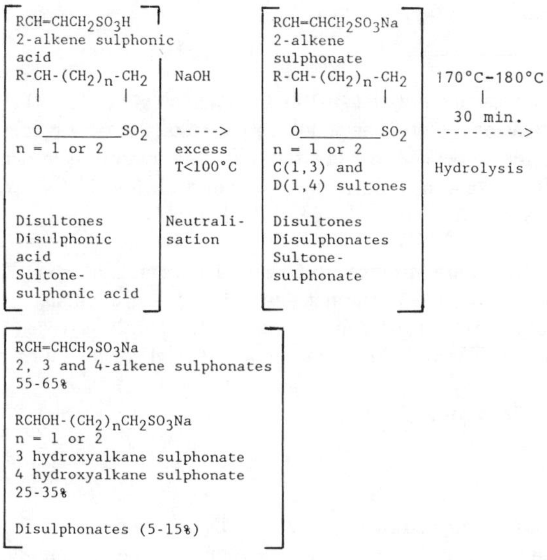

The hydrolysis reaction of C(1,3) and D(1,4) sultones can be described as pseudo first order in the conversion of sultones:

$$-d[C(1,3)]/dt = k_1[C(1,3)] \text{ or } \ln\{[C(1,3)]\ t = 0\} / \{[C(1,3)]\ t=t\} = k_1 t$$

$$-\ d[D(1,4)]/dt = k_2[D(1,4)] \text{ or } \ln \{[D(1,4)]\ t = 0\}/\{[D(1,4)]\ t=t\} = k_2 t$$

In table 8 the hydrolysis rates of sultones are indicated.

Table 8 Hydrolysis rates of C and D sultones at various temperatures

C sultones (1,3) (Gamma sultones) hydrolysis rates

T[°C]	$t_{½}$[s]	$k_1[s^{-1}]$
50	4800	1.44×10^{-4}
60	2100	3.33×10^{-4}
70	1200	5.78×10^{-4}
130	29	2.4×10^{-2}
150	10.5	6.6×10^{-2}
170	4.6	15.0×10^{-2}

D sultones (1,4) (Delta sultones) hydrolysis rates

T[°C]	$t_{½}$[s]	$k_2[s^{-1}]$
110	4200	1.65×10^{-4}
130	900	7.70×10^{-4}
150	174	4.0×10^{-3}
170	46	15.0×10^{-3}

In practise the hydrolysis is carried out at 170-180°C (5 bar pressure) for 30 minutes in a plug flow type of continuous reactor. Hydrolysis above 180°C results in poor colours and introduces corrosion problems. It is recommended that neutralisation and hydrolysis are carried out at different temperatures. By keeping neutralisation temperatures below 100°C, paste viscosities are depressed and this permits a higher solids charge to the hydrolyser.

The combination of neutralisation/hydrolysis gives a product containing about 55-65% alkene sulphonates and 25-35% hydroxy alkane sulphonates and 5-15% disulphonates. The C(1,3) sultones and D(1,4) sultones have virtually disappeared in the plug flow reactor with a residence time of 30 minutes at 170°C. The ratio between initial and final concentration of delta (1,4) sultones will be:

$$\ln Ci/Cf = k_2t = 15.0 \times 10^{-3} \times 30 \times 60$$

$$Ci/Cf \text{ is appr. } 10^{11}$$

i.e. the initial concentration is reduced by a factor 10^{11}. D(1,4) sultones are found in quantities smaller than 10 mg/kg and C(1,3) sultones are not detectable in the end product.

4.3.5 Alpha sulphonated fatty acid methylester sulphonic acid neutralisation and bleaching

Because the acid product has darkened during digestion at 90°C for 30 minutes, two bleaching steps are required. The first bleach occurs prior to neutralisation and is accomplished by adding 1-1.5% (active/active basis) hydrogen peroxide (40% in water) to the sulphonic acid. The peroxide bleaching step continues for 20 minutes at a temperature of 60-70°C. Some hydrolysis of the methyl ester linkage occurs during this bleaching. As a result the final product contains 5-10% alpha sulphonated fatty acid (SFAS). The bleached product is then neutralised at 45°C and the pH is carefully controlled (pH 7.5 to 9) to reduce further ester hydrolysis. A final bleach step using 0.5-1.5% (active/active basis) sodium hypochlorite solution (13% in water) is required to yield an acceptably light product. This bleaching is done at 50-70°C and continues for approximately two hours. Typical product is 30% active. It has an unreacted feedstock (free oil) content of 0.6%, contains 5% disodium sulphonate and has a colour of approximately 50-70 Klett (5% solution, 40mm cell).

Structures for the three sulphonates present in, for example, alpha-sulphomethyl tallowate and for sodium methyl sulphate:

(i)

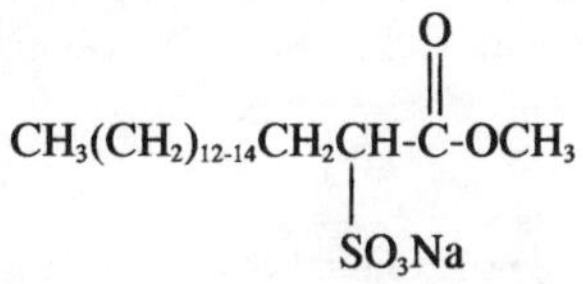

monosodium alpha-sulphomethyl tallowate

(ii)

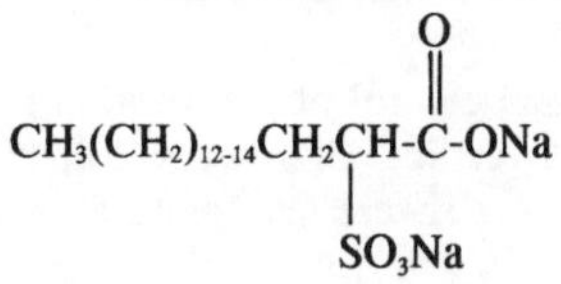

disodium alpha-sulpho tallowate

(iii)

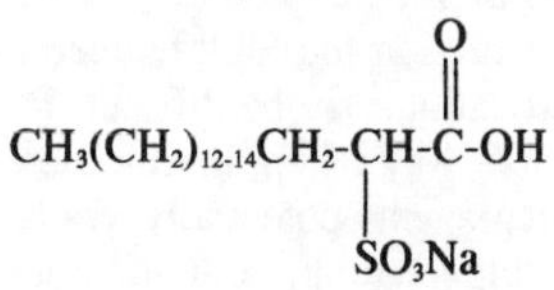

monosodium alpha-sulpho tallow acid

(iv)

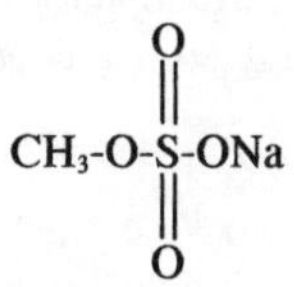

sodium methyl sulphate

4.4 Bleaching chemistry

Linear alkylbenzene sulphonic acid can be made in most types of sulphonation reactor systems and can be stored in the acid form, without serious colour formation. Alcohol sulphates and alcohol ethoxy sulphates can be made without serious colour problems in falling-film sulphonation reactors, provided olefinic impurities are absent, but their colour deteriorates badly (as does conversion also) on storage in the acid form. Alpha-olefin sulphonates require careful control of sulphonation conditions to avoid serious colour formation.

In practice the sulphonation conditions required for FAMES are much more severe than for the other materials, with the result that FAMES poses the worst colour problems. Branched alkylbenzenes may require bleaching notably from long residence-time reactor systems (cascade of continuous stirred tank reactors).

For all the above feedstocks, the initial hydride abstraction (C=C formation) reaction leading to colour has a higher activation energy than the main sulph(on)ation reaction at reaction temperatures commonly used. It follows that, for any feedstock, colour formation will be most serious when:

(i) the sulphonation temperature is not well-controlled;
(ii) conversion is close to completion in the presence of residual SO_3, so that because of the low concentration of unreacted organic feedstock the hydride abstraction leading to C=C bonds becomes more competitive with sulphonation;
(iii) when there is residual SO_3 still present after conversion is completed.

Thus the basic principles for minimising colour are good control of the reaction temperature and use of the minimum of SO_3 necessary for complete conversion. Olefins are colour precursors in any feedstock which is sulphonated under conditions more severe than those appropriate for olefin sulphonation.

The stringent specifications of organic feedstocks and the efficient use of current sulphonation reactor systems has largely solved the colour problems. Significant colour problems can still exist for FAMES and (branched) sulphonic acids which are stored at too high temperatures (above 40-45°C). Consequently, in some instances a colour specification may be difficult to fulfil without a further processing step. The most efficient and convenient way of improving the colour of the sulphonate is to bleach it. There is a variety of bleaching agents potentially available for this purpose. However, only two have any importance for sulphonate bleaching: sodium hypochlorite and hydrogen peroxide. Bleaching is achieved by an agent which 'disrupts' the conjugated C=C bonds, either by reaction with one of the conjugated double bonds or by oxidation and/or reduction of the chromophore, to give extensive breakdown of the chromophoric molecule. Generally, bleaching reactions are not reversible and this is especially so if a double bond has been destroyed.

Sodium hypochlorite is a more convenient and efficient bleach than hydrogen peroxide. However chlorine-based bleach can suffer from the drawback that potentially biologically active material, i.e. sensitisers, may be generated when it is used for bleaching alcohol ethoxy sulphates and alpha-olefin sulphonates.

As an alternative bleach, hydrogen peroxide may be used, but is less cost-efficient and can cause process control difficulties due to excess foaming because of liberation of oxygen during bleaching.

Typical conditions for hypochlorite bleaching are:

Bleaching temperature	40°C
Bleaching time	0.5 - 1 hour
Level of hypochlorite on AD as NaOCl	0.1% - 0.2%
NaOCl concentration (wt%)	14 - 15%

Often the concentration of hypochlorite is expressed as "available chlorine", which is approximately the same as the NaOCl concentration since their molecular weights are about the same, respectively 74.5 and 71. "Available Chlorine" is defined as the measure of the oxidising power of the chlorine present in solution as hypochlorite.

4.5 Exhaust gas treatment chemistry

Unconverted SO_2, unreacted SO_3 (partly as sulphuric acid mist) and entrained organic acid mist droplets are present in the exhaust gas and are the main potential atmospheric pollutants. The sulphur dioxide is due mainly to the incomplete conversion of SO_2 to SO_3. A small proportion of SO_2 may arise from the sulphonation reaction itself but generally the amount of SO_2 in the waste gas is a direct function of SO_2 converter efficiency.

Attention must be given to minimising SO_2 and acid mist pollution on plant start-up, since it takes about two hours to reach a steady state and during this time the $SO_2 \rightarrow SO_3$ conversion increases only slowly from 80% to a level of 98 %.

The other main gaseous pollutant which may be discharged to the atmosphere is sulphuric acid/oleum mist. This is formed mainly from a reaction between non-reacted SO_3 gas and moisture still present in the dried process air and/or derived from the combustion of hydrocarbons commonly present as impurities in the sulphur. This means that the amount of acid formed depends on the efficiency of the air drying system (dewpoint) and the hydrocarbon content of sulphur.

The third pollutant is organic mist which comes directly form entrainment in the reactor system. The amount of entrained organic material is dependent on feedstocks and actual reaction conditions. An order of magnitude value would be 0.2% w/w entrainment on organic feedstock intake.
For the cleaning of the effluent gas various systems are in use. The most common system consists of the combination of an electrostatic precipitator (ESP) to trap sulphuric acid droplets and organic acid mist, followed by SO_2 removal in diluted caustic:

$$SO_2 + 2NaOH \rightarrow Na_2SO_3 + H_2O$$

Small amounts of gaseous SO_3 will also be trapped:

$$SO_3 + 2NaOH \rightarrow Na_2SO_4 + H_2O$$

To avoid any crystallisation of Na_2SO_3 from the caustic solution the max concentration of $Na_2SO_3 + Na_2SO_4$ should be about 10% w/w. The pH of the reaction mixture is kept at a value of 9-10.

4.6. Bibliography

1. Shreve's Chemical Process Industries. 5th edition (1986) Austin G.E., McGraw Hill Book, New York. Chapter 19. "Sulfur and sulfuric acid". pp 320-345.

2. Monsanto Enviro Chem. Bulletin. Vanadium catalysts, for contact sulfuric acid plants.
1) Storage, Handling and Installation of Vanadium Catalysts. 2) Instructions for screening of Vanadium Catalysts.

3. Moretti G.F. (1989). "Main chemical reactions and side reactions in SO_3/air sulphonation of original feedstocks like Alkyl Benzenes, Primary Alcohols, Primary Alcoholethers, Alpha-Olefins and Fatty Acid Methyl Esters", paper presented at Practical Sulphonation Seminar, April 1989 - The Hague, The Netherlands. Center for Professional Advancement, P.O. Box H, East Brunswick, New Jersey 08816, USA.

4. Gilbert E.E. (1965) "Sulphonation and related reactions". Interscience Publishers, New York.

5. Edwards B.E. (1976). "The mechanism of sulphonation and sulphation". Surfactant Science Series, Anionic Surfactants part 1, Vol. 7, p 111, edited by Linfield W.M. Dekker, New York.

6. Knaggs E.A. et al. (1978). "Sulfonation and sulfation". Kirk-Othmer Encyclopedia of Chemical Technology, 3rd edition, Vol 22, Wiley Interscience, New York.

7. Falbe J. (editor) (1986). "Surfactants in consumer products. Theory, technology and applications". Springer Verlag, Berlin.
Section 3.2.1, Anionic Surfactants, pp 53-79.

8. Mazzoni SpA, Busto Arsizio, Italy (1988). "Mazzoni SO_3/air sulph(on)ation processing technology". Booklet, pp 26-31, Chemistry.

9. Cerfontain H. et al. (1980). J. Chem. Soc. Perkin Trans. 2, p 28, and references sited therein. (Topic: Alkyl Benzene Sulphonation Chemistry).

10. Efimov V.T. et al. (1980). Russian J. of Applied Chem. 53 p 1847 (Topic: Primary Alcohol Sulphation Chemistry).

11. Harlbert R.C. (Shell) (1967) Soap & Chemical Specialities, p 122 (Topic: Primary Alcohol Sulphation Chemistry).

12. Whyte D.D. (1955) "Instantaneous sulphation of fatty alcohols", The JAOCS, June 1955, Vol. 32, no. 6, p 313.

13. The Chemithon Corporation, Seattle, USA (1988). "Application of pilot studies to minimizing 1-4 dioxane production during sulfation of ethoxylated alcohols".

14. Moretti G.F. and Adami I. (1988). R&D Dept. Ballestra SpA, Milan, Italy. "Sulphation of ethoxylated alcohols in multitube film reactor: Product quality and reaction control for low dioxane content". Paper presented at the 2nd World Surfactant Congress, Paris 24-27 May 1988.

15. Foster N.C. (1989) "Alpha-Olefin sulfonation and sulfonates". Paper presented at Practical Sulphonation Seminar. See reference 3.

16. Nagayama M. et al. (1973). Chem. Commun. 1973 p 841. (Topic: Alpha Olefin Sulphonation).

17. Nagayama M. et al. (1974). Bull. Chem. Soc. Japan 47 p 2158. (Topic: Alpha-Olefin Sulphonation).

18. Roberts D.W. et al., J. Chem. Soc. Perkin 2 p 389. (Topic: Alpha-Olefin Sulphonation).

19. Miyauchi T. et al. (1971). J. Chem. Eng. Japan 4 p 44. (Topic: Alpha-Olefin Sulphonation).

20. Yih S.M., Seagrave R.C. (1980). Chem. Eng. Sci. 33, pp 1581-1587. (Topic: Fatty Acid Methylester Sulphonation).

21. Henstoil W.H., Hanratty J.J. (1976). AIChE 22, no.6, p 990. (Topic: Fatty Acid Methylester Sulphonation).

22. Kapur B.L. et al. (1978). JAOCS Vol. 55, no.6, pp 549-557. (Topic: Alpha Sulfo Fatty Acid Methylester).

23. Bonner G.A., Tuvell M.E., Ethyl Corporation USA. "Sulfonated Me. esters, their synthesis, composition & application".

24. Adami I. et al., R&D Dept., Ballestra SpA, Italy. "Fatty Acid Methylester, a universal raw material for detergent manufacture".

25. Lohr J.W. (1958) JAOCS, p 532. (Topic: Neutralisation of reaction product of primary alcohol sulphation).

26. Batts B.D. (1966) "Hydrolysis in aqueous acid solutions", J. Chem. Soc. (b), p 551.

27. Falbe J. (editor) (1986). "Surfactants in consumer products etc., see ref. 7. Section 5.1.2.3. Bleaching Systems, pp 272-274.

28. Roberts D.W. et al. (1988). (Topic: Ester Sulphonation), Proc. 2nd World Surfactants Congress, Paris, 24-27 May 1988, vol. 2, pp 38-41.

29. Roberts D.W. et al. (1986). (Topic: PAS-acid decomposition kinetics, J.A.C.S. 63, pp 799-803.

30. Roberts D.W. et al. (1983). (Topic: Sultone by-product in anionic surfactants, formation and allergenic properties), Tenside 20, pp 109-111.

31. Roberts D.W. et al. (1990). (Topic: AOS/bleach interactions and unsaturated sultone formation), Tenside 27, pp 82-86.

32. Roberts D.W. personal communications, 1990.

5 PROCESSING EQUIPMENT

5.1 Introduction

In this section the processing equipment as supplied by Ballestra, Chemithon, Meccaniche Moderne and Mazzoni and Lion is described. The differences in SO_3/air raising technology between these suppliers are marginal and therefore the discussion of SO_3/air gas raising equipment is not subdivided between suppliers, but differences are highlighted. The sulphonation reactor technology and neutralisation equipment differ significantly between the four suppliers and therefore they will be discussed separately. Exhaust gas cleaning and effluent treatment are sufficiently similar to be described in general terms with specific remarks where significant differences between the four suppliers occur.

5.2 SO_3/air generation

Figure 7 outlines in a block diagram, the various elements which comprise SO_3/air generation. The following steps can be distinguished:

- Sulphur melting and refining (5.2.1)
- Air drying equipment (5.2.2)
- Sulphur burners (5.2.3)
- SO_2/air coolers and hot gas filters (5.2.4)
- $SO_2 \rightarrow SO_3$ converter tower (5.2.5)
- SO_3 absorbers (5.2.6)
- SO_3/air coolers, oleum collection and mist filtration (5.2.7)
- SO_3/air gas splitting (5.2.8)

Conclusions are presented in section 5.2.9.

Fig. 7 Sulphur burning plant and SO_2 to SO_3 conversion plant: block diagram

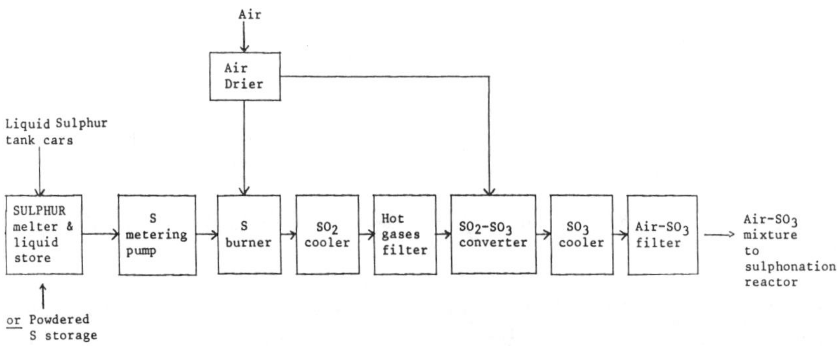

5.2.1 Sulphur melting and refining (see also 3.1)

In many European countries and in North America, sulphur can be delivered in liquid form and stored on site in heated storage tanks. Sulphur melting is required when sulphur is delivered in solid form which is the case in almost all African countries (Sulphonation plants in Nigeria, Ghana, Kenya), many Asian countries (India, Thailand, Philippines, Indonesia, Malaysia) and in a few Latin America countries (Brazil, El Salvador, Uruguay).

Sulphur storage and handling procedures can be found in section 3.1.

When located in a building, the sulphur melting tank should be equipped with a suction hood (possible release of H_2S and SO_2). The bottom of the entire melting vessel should be provided with steam-heating facilities (e.g. half-pipes welded to the bottom) to be able to heat up bottom residues during periodic tank cleaning. Heating should not be in use during normal operation.

Contamination of the sulphur increases the need for maintenance (e.g. corrosion of steam coils) and cleaning of the melting vessel. If sulphur is clean and dry cleaning may not be necessary for several years.

It is recommended to have the bottom of the vessel slightly sloping and to have a bottom discharge (wide flange) to facilitate cleaning. The viscosity of molten sulphur varies with temperature (3.1). Experience has shown that the optimum safe handling temperature is between 130°C and 150°C and the chosen temperature should be maintained as accurately as possible to avoid dosing fluctuations due to significant viscosity changes.

The mild steel melting vessel is normally divided into three different sections:

A) The melting zone

This is the area where solid sulphur is charged and where melting takes place. This section should be separated from the next section by means of a coarse mesh stainless steel grid.

At the sulphur feed point, a steel grid with 6" square openings is installed over one end of the melting compartments as a safety device and to screen out large lumps of tramp material.
A small hopper can be built around the grid to permit accumulation of a sulphur pile above the grid and to provide additional free space to prevent foaming over. As the sulphur melts under the grid, the pile gradually sinks into the melter. Regular re-charging of solid sulphur would require only a minor amount of the operator's time on each shift.

Continuous melting has proved more satisfactory than intermittent or batch melting. Intermittent melting results in wide fluctuations in the level of molten sulphur. This causes increased steam coil corrosion, because the coils are alternately exposed to sulphur and air.

The steam coil banks are usually designed and installed so that in case of failure of any single unit it is a simple matter to remove the faulty unit without shutting down the entire melting system. The melter is usually provided with removable steel or aluminium covers to exclude dust and reduce heat losses.

B) The decantation zone

Molten sulphur overflows from the melting compartment into the decantation zone, subdivided into several compartments which is by baffles. This permits more efficient settling and skimming of floating debris.

C) The final zone

This section, from which a pump takes the sulphur, is separated from the decanting/settling zone by means of a bottom baffle-plate. From the final zone, the sulphur can be sent either:

(i) straight to the process of sulphur combustion (sulphur burner) via a suitable filter system (two in parallel) and a proportioning pump. Recommended pumps are Bran and Lübbe or OMG (Gellarates), for Ballestra and Milton Roy (Chemithon). The sulphur feed pumps are piston-type with the body and valves jacketed for heating with steam at 4 bar (150°C) and are driven by electric motors; or

(ii) to a sulphur storage day tank via a suitable filter system (two in parallel) from where the sulphur can be dosed to the sulphur burner.

In this case, the sulphur burner is less dependent upon the sulphur melting process. The day or shift-tank can be placed on load cells so that the metering pump for sulphur can be checked by the loss-in-weight principle.

Nowadays in-line sulphur mass-flow measurement is recommended. For sulphur transfer to an intermediate day-tank various types of pump are used: submerged centrifugal pumps (Lewis type), horizontal, steam-jacketed centrifugal pumps or Viking gear pumps. The piston-type positive displacement metering pumps of the Bran and Lübbe or Milton Roy type may be replaced in the near future by gear-pump mass flowmeter (Micromotion) combinations.

The primary purpose of sulphur filtration is to remove solid contaminants or "ash" that might clog the catalyst mass in the SO_2-SO_3 converter tower. The removal of solid contaminants also decreases maintenance costs and improves the operation of burners, pumps, valves, meters and other sulphur processing equipment. Filtration is more effective than sedimentation because ash is often finely divided and very slow in settling. It also eliminates the large holding and storage tanks required for effective settling. However, in cases where the only sulphur available has a high ash content, sedimentation may prove more effective to avoid too frequent filter cleaning.

Sulphur, as mined, contains only very small amounts of ash, usually less than 0.01%. Sulphur "ash" content can rise to 0.3% or more during or after shipment. The impurities introduced in transit and at plant-site storage consists of wood splinters, rust scale, tramp iron, fly ash, sand, dirt, dust and ores remaining from cargoes previously carried in transportation. The two basket filters (one in stand-by position) for molten sulphur, inserted between molten sulphur tank and metering pump, contain a filtering element made up of a stainless steel net.

Liquid sulphur processed from natural gas and mineral oil resources does not present these problems. Nevertheless filtration as outlined above is performed between store tanks and metering pump or day tanks and metering pump.

It is of paramount importance to use high purity grade sulphur for sulphonation and to verify that this grade is always supplied. Variations in total sulphur dosing cannot be tolerated as this will have direct consequences for product quality.
Cheap grade sulphur is false economy.
If the supplier cannot meet the specified quality, contact an engineer of the sulphonation equipment supplier. Poor quality sulphur can be upgraded by a refining process. Molten sulphur is recycled over a pre-coated filter, where organic material, extraneous solids and inorganic salts (from acid sulphur, H_2SO_4 and H_2S, neutralised with ammonium bicarbonate) are trapped in the filter.

The pre-coated filter can operate for a period of two to four weeks, depending on the level of impurities in sulphur. The equipment supplier can help you either to find a sulphur refining operation in the region to do the job or supply a sulphur upgrading unit.

5.2.2 Air drying equipment (see also fig. 8)

A large quantity of dry air is required to provide the oxygen for the conversion of sulphur to sulphur trioxide and to dilute the SO_3 to a concentration of 4-7% v/v SO_3 in air for use in a sulphonation reactor. Assuming 100% conversion and 1 kg sulphur:

dry air Nm^3/h ($0^\circ C$, 1 bar)	dry air kg/h	$\% SO_3$ in air
17.85	23.03	4
14.35	18.51	5
12.02	15.50	6
10.35	13.35	7

Calculation: $Nm^3/h = (kg/h \text{ of sulphur}) (70/\%SO_3 + 0.35)$

Moisture in the air gives rise to the formation of sulphuric acid and oleum, leading to corrosion throughout the plant and causing inferior product quality, in terms of colour and other undesired by-products (e.g. dioxane in alcohol ethoxylate sulphates) if condensed oleum reaches the reactor.
Therefore moisture in the process air entering the sulphonation plant must be kept low with an upper limit of 0.01 g/m^3. This corresponds approximately with a dewpoint of the dried air of $-60^\circ C$, the dewpoint being the temperature at which moisture will condense out of the air.

Dewpoints can be measured manually with simple dewpoint apparatus but the use of an in-line dewpoint meter guarantees continuous monitoring of the moisture levels in air. The recommended instrument for in-line measurements is the Shaw hygrometer type HD.

The importance of the air dewpoint on sulphuric acid/oleum formation in the gas raising plant can be seen from the following table:

Table 9 Kg 20% oleum per 24 h for a plant with a capacity of 1 ton LAS/h at various process air dewpoints

Dewpoint of process air °C	gH_2O/kg dry air	7% SO_3 in air	4% SO_3 in air
- 20	0.634	138	238
- 30	0.234	51	88
- 40	0.079	17	30
- 50	0.024	5	9
- 60	0.0066	1.5	2.5
- 70	0.0016	0.5	0.5

Example: calculation for process air dewpoint -20°C, 7% SO_3 in air:

kg H_2O/kg dry air x kg dry air/h x 24 x 100/kg water per 100 kg 20% oleum =

$0.634 \times 1335 \times 24 \times 10^{-3} \times 100/14.7 = 138$ kg 20% oleum.

To meet the above specification of process air with a dewpoint of at least -60°C, air drying equipment must be installed.

The size of the equipment depends on the local atmospheric conditions. When specifying an air drying plant, detailed data on the local climate must be given to the sulphonation equipment supplier, so that the correct air drying plant is obtained. A typical air drying plant is depicted in figure 8.

The air drawn in from the atmosphere is filtered and compressed through the compressor which produces the required pressure to overcome the resistance of the total plant, including the exhaust gas cleaning system. The pressure at the compressor exit varies with the type of sulphonation plant but is in the range 0.5 - 1 bar gauge.
The air should not be drawn from the blower room but from outside and filtered through a correctly designed and dimensioned filter. Proper air filtering is an important detail.

Ensure adequate ventilation to dissipate heat in the compressor room. The compressor, normally Roots-type, should be of the continuous, dry operation type to eliminate any oil carry-over in the process air. The process air flow is pulsation free and at a constant pressure. After the compressor the air is cooled in a heat exchanger with cooling water to remove the air compression heat effect.

The process air is then further cooled to about 3-5 °C in a heat exchanger utilising a refrigerated glycol water mixture. The cooling medium must be controlled such that ice formation on the heat exchanger tubes is avoided. The glycol-water cooling system therefore consists of two loops: one cools the glycol/water mixture with a normal Freon expansion chiller, and the other cools process air with an adequate amount of the recirculation glycol/water mixture. The process air, which should not exceed a temperature of 5°C, carries condensed water which should be effectively removed in a water separator.

When, for example, ambient air of 20°C with an humidity of 70-80% containing about 10 g H_2O/kg dry air is cooled to 5°C, corresponding with a saturation content of about 5 g H_2O/kg dry air, at least 5 g water per kg dry air condenses out of the process air. In other words, the chiller in this example removes 50% of the water initially present in the process air. The water that has to be separated from the process air stream would amount to about 50 litres per hour assuming, for example, a 5 ton AD/h plant using 5% SO_3 in air.

The cooled, compressed air is then passed through a volumetric measuring device (orifice air meter) to two parallel absorbers operating in sequence i.e. one on air drying and the other on regeneration. The intermittent operation of the two parallel driers is completely automatic and can be linked to any control system. Although the operating sequence is batchwise, process air flow pressure and dewpoint of dried air should be absolutely constant. The Mazzoni design has a connection between the two driers for pressure equalization after regeneration, which ensures a pulse-free bed change over.

The regeneration air for the desiccant (silica gel, activated Alumina) may be heated for the first start-up by means of a steam coil heater, but during normal plant operation heating is accomplished through the hot air recovery from the SO_3/air coolers. It is necessary to check on a regular basis that the hot air for regeneration is not contaminated with SO_2/SO_3 from leaks in the system. If the reliability of the SO_2 and SO_3 air coolers is considered to be an issue or if indirect-fired boilers are used, then it is often more appropriate always to use steam for regeneration of the driers. Ensure that the regeneration cycle time is such that the required heating temperature is reached, the maximum temperature not exceeded, and that proper cooling of the bed after regeneration to the recommended temperature is adhered to (table 10: temperature cycle during regeneration of desiccant). Check the correct operation of the dewpoint meter on a regular basis.

Silica gel is mostly used in the absorption tower to dry the air from a dewpoint of 3-5°C to a dewpoint of as low as -60°C.

Table 10 Temperature cycle during regeneration of desiccant (silica gel), 1500 kg ABS/h, 5% SO_3 in air, Ballestra plant, temperature regeneration with hot air 160°C

Time (h)	Outlet temp. [°C] air, leaving silica gel drying bed	Cycle
0	5	Bed temperature before regeneration, start of introduction of hot air
0.5	15	Heating of silica gel bed
1.0	30	Heating of silica gel bed
1.5	105	Major part of absorbed water removed
2.0	120	Further water removal
5.0	135	End of hot air cycle, start of cooling with cold air (30°C)
6.0	100	Cooling cycle
7.0	60	Cooling cycle cont'd
8.0	45	Cooling cycle cont'd
9.0	32	End cooling cycle, bed ready for use

Figure 8 Airdrying equipment

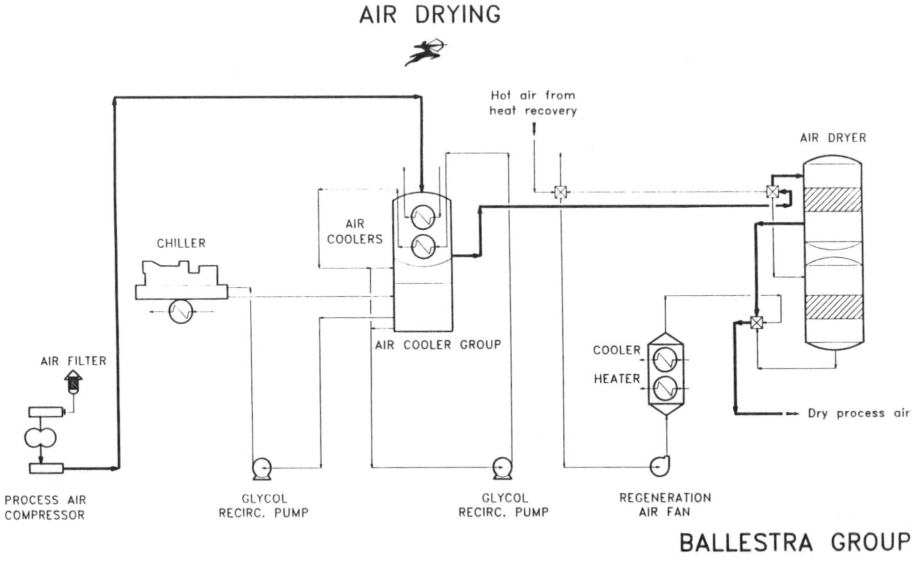

5.2.3 Sulphur burners (fig. 9)

Sulphur burners exist in two designs: vertical and horizontal. The vertical burner operating co-currently or countercurrently, consists of a vertical steel cylinder lined with insulating brick and fire resistant brick. The upper section contains firebrick chequer work supported on a refractory arch. The lower section, which is under the arch, serves as a supplementary combustion space. Liquid sulphur and combustion air enter at the top; the sulphur vapourises and burns as it runs down through the chequer-work of firebrick. Any vapour not burned in the chequer-work burns in the lower combustion space. The hot sulphur dioxide containing gas leaves the burner at the side of the bottom portion. A countercurrently operated vertical design is used by Ballestra, while Mazzoni and Meccaniche Moderne use a cocurrent vertical burner. In the Ballestra design molten sulphur enters at the side of the furnace via a gun pipe fabricated in high-temperature alloy. The gun pipe must be inspected at annual intervals. In the MM and Mazzoni design sulphur is introduced in the top of the furnace through a pipe. The process air is pre-heated to 150 °C and enters around the sulphur pipe outlet in the top of the burner.

The horizontal burner consists of a horizontal steel cylinder lined with firebrick and insulating brick. A firebrick chequer is located in the front compartment. Combustion air enters at the end of the burner below the firebrick chequer. The rest of the burner is subdivided into compartments with over-and-under air pass baffles, which help complete the combustion of the vapours. The flow of liquid sulphur and air oppose each other in the chequer. The hot brick work vapourises and ignites the sulphur as it flows down and the combustion of the sulphur vapour is completed in the after compartments. Chemithon supply the horizontal type sulphur burner.

Corrosion or "hot spots" are not usually a problem. Burners have operated for years with no extensive repairs to lining or shell. Assuming good workmanship during construction, the effective curing of the refractory lining during commissioning and the use of good quality clean sulphur the life of the burner depends on the number of shutdowns, the care taken when starting up and the temperatures used. The chequer-work is usually the first part to fail, but should last a number of years. Neither burner nor chequer-work need frequent cleaning provided that good quality, clean sulphur is burnt.

The sulphur ignition temperature is about 250 °C (see 4.1.1.). A cold burner therefore needs preheating with combustion gas from oil or gas for 2-3 hours and to vent the wet combustion gases from a "starting stack" on the burner. After the burner brickwork is heated the preheating is stopped and the starting stack is blanked.

Ballestra do not preheat the furnace but electrically ignite the sulphur flow to the furnace. It is important to follow exactly the recommended operating procedure and particularly to ensure that the electric elements are drawing the specified current and the supplementary combustion air is at the correct pressure and flowrate (small rotameter). On start-up the temperature of the furnace rises to the auto-ignition temperature of sulphur in approximately 3-5 minutes at which point the ignition procedure can be terminated.

Correct operation of the burner is essential for successful functioning of the total sulphonation plant. Unburnt sulphur may evaporate and will pass upstream with the process gas causing sulphur sublimation deposits in filters and catalyst tower ("yellow fever"). Subsequent local burning of sulphur can cause severe damage to equipment such as the SO_2 cooler, the SO_3 filter, the catalyst support grids and other parts of the converter tower. This will also cause an irregular level of SO_2 and therefore SO_3 in the process gas leading to poor sulphonation product quality. Too low temperatures in the sulphur burner may cause irregular burning resulting in varying levels of SO_2/SO_3 in the sulphonation process gas. An outlet temperature of the burner between 600-650°C should be maintained i.e. the SO_2 level in the gas leaving the burner should be in the order of 6-7 vol% (see 4.1.1.).

The air inlet flow, temperature and pressure must be controlled and fitted with an interlock to shut off the sulphur feed if there is insufficient air flow. Molten sulphur is metered through a metering pump or a mass flowmeter together with the amount of pre-set air, corresponding to the SO_2 concentration desired at the furnace outlet.

Figure 9 Sulphur burner

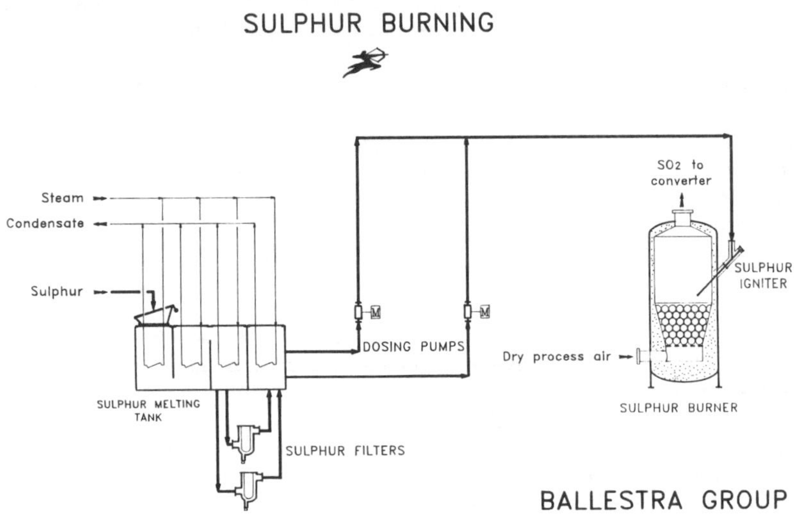

5.2.4 SO_2/air coolers and hotgas filters

The SO_2/air gas of 600-650°C has to be cooled to about 450°C. Waste heat boilers can be used to raise saturated steam of 15 bar (195°C). A waste heat boiler has the advantage of quickly reducing the gas temperature and supplying steam to the factory steam network. 3 kg steam can be raised per kg sulphur burnt. Indirect systems are intrinsically safer.

The commoner and preferred alternative is to use air coolers of simple design. The heated ambient air up to temperatures of 250-400°C is mixed with the quench air of the NSD spray tower furnace. This direct application of heat is simple and cheap provided that the distance between the sulphonation operation and the NSD spray tower is no more than 100 m. A lagged pipeline connects the sulphonation plant and the spray tower furnace.

Energy savings in the order of 10% can be obtained in the spray-drying plant depending on plant capacities of sulphonation and spray-drying and the overlap period of running both operations. Sulphonation should preferably run fully continuously, while NSD spray drying plants may not operate during weekends and may even be operated on a one or two shift mode during normal working days. The total effective waste heat amounts to 10^6 kJ per 100 kg S burnt.

The Ballestra group uses a combination of U-tube cooler and a shell-and-tube heat exchanger for the cooling of SO_2/air with ambient air. MM and Mazzoni supply a vertical double-hairpin heat exchanger (a so-called trombone cooler) with air cooling.

The hot SO_2/air gas flows through the stainless steel 316 inner pipe and the ambient cooling air through the mild steel outer pipe. The first part of the hairpin cooler is used to preheat the dried process air before it enters the top of the sulphur burner. The Chemiton SO_2/air-ambient cooler is a simple single horizontal hairpin double-tube heat exchanger.

SO_2/air leaving the burner will contain traces of "ash" and "dust" stemming from the refractory material lining the sulphur burner. Therefore the cooled SO_2/air is passed through a hot gas filter before entering the converter tower.

The hot gas filter is made up of a cylindrical vertical stainless steel casing, housing the filtering element. The filtering material is made up of a layer of refractory granules, 6-8 mm diameter, about 400 mm high, supported by a stainless steel grid and covered with a layer of quartz pebbles. On top of this filter bed is another layer of quartz pebbles to distribute the SO_2/gas evenly over the cross-section of the filter. The SO_2/air enters the upper part of the vessel, passes through the filtering bed and leaves from the bottom of the vessel. This filter should prevent dust from collecting in the converter tower top catalyst bed. Dust collection in the converter tower causes an increased pressure drop and if a sufficient part of the surface of the catalyst is covered, then $SO_2 \rightarrow SO_3$ conversion would drop.

5.2.5 SO_2 - SO_3 converter tower (see figure 10)

The reactor for chemical conversion of sulphur dioxide to sulphur trioxide is designed to maximise the conversion by taking into consideration that

(i) equilibrium is an inverse function of temperature and a direct function of the oxygen to sulphur dioxide ratio.
(ii) the rate of reaction is a direct function of temperature.
(iii) gas composition and amount of catalyst effect the rate of conversion and the kinetics of the reaction. (See section 4.1.2)

The converter tower should be made of stainless steel 316. The catalyst mass on each of the 4 beds is "sandwiched" between two layers of quartz: the top layer ensures proper gas distribution to the underlying catalyst bed and the underlayer prevents catalyst pellets falling through the stainless steel supporting grid. There is a manhole above each bed, used for catalyst inspection and catalyst replacement. Ideally these manholes should be accessible by a permanent scaffolding construction with a ladder and a small platform to each bed position. Thermocouples are located after entry and at the exit of each of the converter beds. The operation of the converter is controlled by the entry temperature of each bed. It should be verified that each thermocouple is correctly positioned and in good condition.

The pressure drop over each bed should be monitored every 6 months, using the pressure tapping points provided.

The optimal entry temperature to the first bed is between 410°C and 445°C. The exit temperature at the first bed should be no more than 610°C because higher temperatures can damage the internal surface of the converter and also reduce the catalyst activity. A high and low temperature alarm is located at the first catalyst bed inlet. Between the first and second bed the reaction gas has to be cooled to 440°C, the entry temperature for the second bed. Heat exchangers with cooling air are normally installed for cooling the gas from the first bed. The heat can be utilised in various ways: for regeneration of the desiccant in the process air drying plant, for steam raising or hot air production for the NSD spray-drying tower furnace.

The temperature rise over the second bed will be considerably less (from 440°C to 490°C). A second heat exchanger between the second and third beds can be installed (Ballestra technology) or fresh dried air can be added for cooling. The former method gives more flexibility in plant operation.

This system offers the advantage of having the option to sulphonate at high SO_3 levels in air, say 6%, which is feasible for linear alkylbenzene feedstocks. The dilution with dry air offers the advantage of saving a heat exchanger (MM, Mazzoni). If a maximum conversion of 92-95% has been achieved in the first two beds and if the plant is not overloaded, the entry temperature to the third and fourth beds that will ensure the maximum yield of 98.5% will be in the range 430-440°C.

At plant shut-downs the catalyst should be purged with dry process air for half an hour to remove the SO_3 absorbed by the catalyst itself. After the cleaning with air, care must be taken that moist air does not enter the converter tower. Therefore all valves must be closed to prevent ambient air coming into contact with the catalyst. The catalyst is hygroscopic especially when it has been used for the $SO_2 \rightarrow SO_3$ conversion. Catalyst handling and sieving instruction are found in Appendix 3.

The converter tower catalyst has to be preheated before plant start-up. A gas or oil-fired preheater (Ballestra) supplies the hot combustion gas which heats the dried process air, using the SO_2 cooler as a preheater. A temperature of 400°C in the catalyst tower is attained after about 3 hours.

MM and Mazzoni prefer an electric preheater in a closed circuit. Dried process air is circulated through the converter tower, hot gas filter and electrical elements. Electrical heating offers the possibility of variable heat input. For example, during weekend stoppage (not recommended!) the electrical preheater can be set to maintain a high temperature in the converter tower and hot gas filter. This results in a quick start-up, which contributes to long catalyst life and consistently high conversion. Moreover, a rapid start-up will increase the actual effective production time.

The problem with electrical preheaters is the danger of severe corrosion of the electrical heating elements in the event of SO_3/air leakage through the isolation valves of the preheating circuit, during regular production. Mazzoni claim improved reliability using incoloy encased heating elements. Chemithon uses a system of electrical heaters mounted on the vessel shells to maintain sulphur burner converter temperatures during shut-downs. This system is not subject to corrosion failures.

Figure 10 SO_2/SO_3 Converter tower

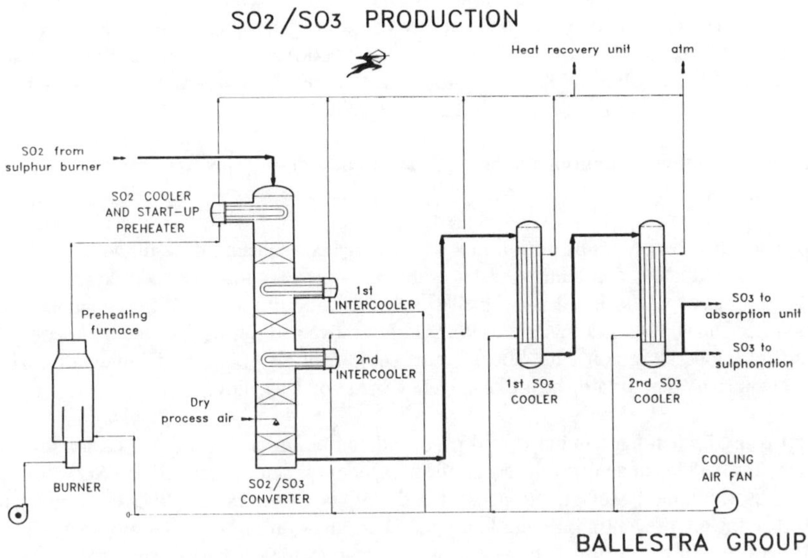

Figure 11 Heat recovery system

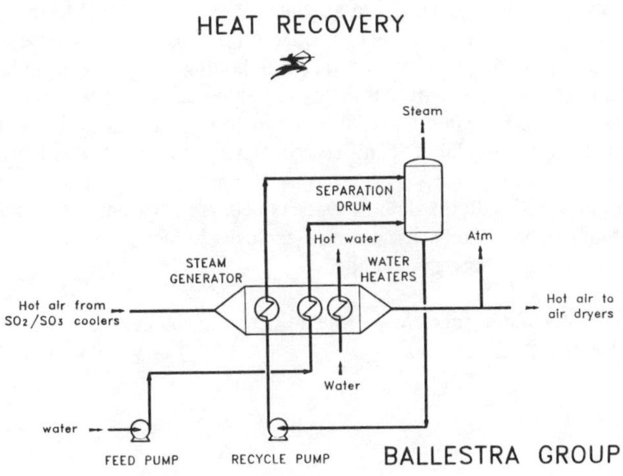

5.2.6 SO_3 absorbers (fig. 12)

After preheating of the plant and initial sulphur burning, the conversion to SO_3 will increase gradually with time. It takes about 0.5 hour before the conversion attains the desired level of 98-98.5%. During this period it is not possible to produce good quality organic acid product and therefore the SO_2/SO_3/air gas flow has to be diverted, without infringing the limits on SO_2/SO_3 levels in the exhaust gas that are emitted to the atmosphere.

There are two well-proven solutions for SO_2/SO_3 absorption during this part of the plant start-up procedures.

The first method involves SO_3 absorption in a scrubber where concentrated sulphuric acid is circulating. A stoichiometric amount of water is dosed to the SO_3 absorber to produce about 98% H_2SO_4. The amount of water dosed is controlled by measuring the in-line H_2SO_4 conductivity. The reaction heat of H_2SO_4/H_2O reacting with SO_3 is removed in a heat exchanger, keeping the circulating acid temperature constant. Special steel equipment e.g. pipes and valves are made of teflon-lined materials. The (plate) heat exchanger is made of Hastelloy.

This system is used when the sulphuric acid produced can be utilised locally or easily sold to third parties. The sulphuric acid produced is of an excellent technical grade. The SO_3 absorber tower is also used in emergency situations so that the SO_3/air flow can be diverted immediately to the absorber tower, preventing a complete plant shut-down procedure. The SO_2 in the partially converted mixture will be absorbed in the caustic scrubber, which is part of the exhaust gas cleaning system. The design of the SO_2 absorber should be adequate to absorb SO_2 effectively during the start-up stage.

The second method involves SO_3 absorption in alkylate that is recirculated through the sulphonation (film) reactor. The sulphonic acid formed returns to the fresh alkylate holding tank which forms part of the circulating system. The alkylate holding tank should be large enough to ensure that the contents are only partially converted to sulphonic acid at the end of the start-up period. The partially converted organic material is dosed from this tank (at a low concentration) into the organic feed to the sulphonation (film) reactor during regular production of organic acid.

This system is preferably used with alkylate feedstocks because sulphonic acid is a stable product. For those sulphonation plants that produce exclusively alcohol sulphates, alcohol ethoxy sulphates etc. an H_2SO_4 absorber is recommended.

The SO_2 in the partially converted mixture will be absorbed in the caustic scrubber, which is part of the exhaust gas cleaning system.

Figure 12 SO_3 absorption by sulphuric acid

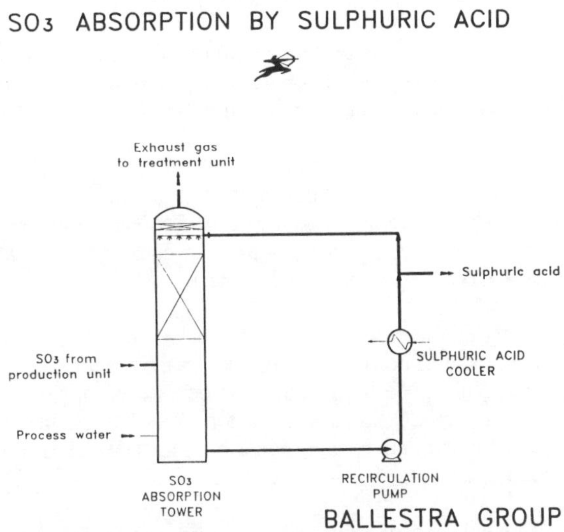

5.2.7 SO_3/air coolers - oleum collection and mist filtration

The gas mixture leaving the converter tower has a temperature of about 430°C, which is far too high to enter the actual sulphonation stage. Therefore, the SO_3/air is cooled with air-cooled shell-and-tube heat exchangers (stainless steel tube bundle) in series. The heated cooling air from the first heat exchanger can be recovered as the temperature is sufficiently high. In subsequent heat exchangers the SO_3/air is cooled to 50°C. A small amount of about 20% oleum condenses out of the gas and is collected in a small tank under the tube bundle of the heat exchangers. The lower the process air dewpoint (see 5.2.2), the smaller is the amount of oleum formed.

The treatment of this collected oleum is described in section 7.3.1. A small amount of oleum mist is formed which has to be removed before the sulphonation stage. Oleum reacts rapidly with organics, leading to poor colour/charring of organic feedstock and by-product formation (e.g. dioxane). Moreover oleum droplets or crystals can cause maldistribution of organic feedstock in the top of the reactor. The installation of a high-efficiency demister is therefore recommended.

The demister may reduce the formation of dioxane during alcohol ethoxylate sulphation and may also reduce corrosion and fouling at the head of the reactor by reducing the exposure to oleum. The containing vessel of the demister should be hot water traced. The liquor drained from the vessel bottom is a solution containing oleum and iron sulphate. The filter should be installed as close as possible to the reactor top. The gas tube from the SO_3 gas generation unit through to the sulphonation reactor should be thermally insulated.

5.2.8 SO_3/air gas splitting (fig. 13)

Whenever simultaneous production of different sulphonation/ sulphation active detergents is required, the SO_3/air flow has to be split across two or more sulphonation reactors. The total

SO_3/air flow required can be produced by one single gas raising plant. SO_3/air gas splitting across various reactors will be beneficial in the following operational examples:

(i) the production of two different types of detergent on two dedicated sulphonation reactors, avoiding large stocks of finished products that would be required if one reactor were installed. The two different actives may require different SO_3 levels in air;

(ii) the combination of producing an active detergent and sulphuric acid at the same time. In some countries, the cost of imported sodium sulphate is so high that in-house production can bring substantial savings. The sulphuric acid produced can be mixed with sulphonic acid and subsequently neutralised, producing a paste of equal AD and sodium sulphate strength (approx. 25% AD, 25% Na_2SO_4, 50% water).

Figure 13 indicates a gas splitting system where two different organic feedstocks are sulph(on)ated on two parallel reactors, however the SO_3 concentration for reactor 'B' is reduced by means of extra dilution air. This mode of operation can be used when reactor 'A' produces sulphonic acid using a relatively high SO_3 in air and reactor 'B' is fed with a more sensitive organic feedstock, for example alcohol ethoxylate, requiring a lower level of SO_3 in air to produce a good quality active detergent.

Impurities (condensation of oleum mist, $H_2S_2O_7$) and "dust" have so far been the main reason for manual operation of originally designed automatic gas-splitting systems. With the falling-film sulphonation reactor technology and the requirements for high-quality products with a minimum of undesired by-products (e.g. dioxane), a proper and stable operation is impossible without correctly designed filters throughout the SO_3 gas-raising plant.

5.2.9 Conclusions

Consistently produced, clean and dry SO_3/air is of paramount importance to obtain good quality sulphonation products at high yields in terms of raw material costs and capital employed. Therefore:

- purchase good quality sulphur. Cheap sulphur is false economy;
- ensure proper drying of process air to a dewpoint of -60°C/-70°C;
- a steady sulphur burner temperature at the exit (± 5°C) should be maintained. Recommended ignition procedures must be followed to avoid accumulation of sulphur in the furnace. Avoid "yellow fever"!
- monitor temperature rise over each bed in the converter tower and check pressure drop over each bed. Manufacturers' procedures must be followed when handling catalyst;
- check oleum collection from last SO_3/air cooler;
- install an efficient demister to prevent oleum reaching the organic feed distributor in the reactor.

Figure 13 SO_3/air gas splitting

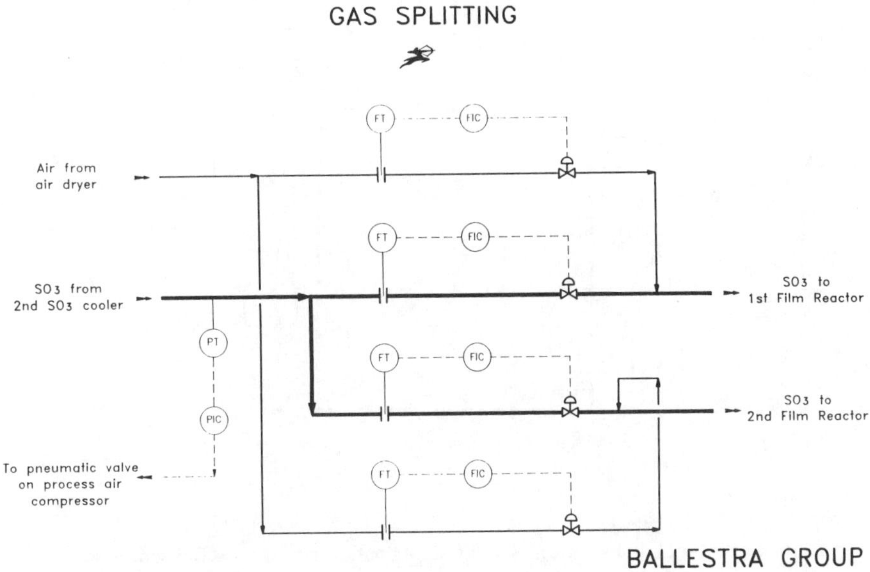

5.3 SO_3 generation from liquid SO_3

An alternative to sulphur burning and subsequent SO_2 conversion to SO_3 is the use of stabilised liquid SO_3. The use of stabilised liquid SO_3 is restricted by the proximity of a supplier. This chemical should be transported and handled with great care. A simplified flowsheet is presented in figure 14. The stabilised liquid SO_3 is transferred by a membrane metering pump from the day tank through a filter and into the vaporiser. The gaseous SO_3 thus obtained is filtered to remove any remaining mist and then mixed with dried process air (dewpoint -60 °C) to the required SO_3 level.

The advantages of the stabilised SO_3 route are the following:

- lower plant capital costs compared with sulphur-based SO_3/air sulphonation;
- reduced plant start-up/shutdown time spans and overall easier operation compared with sulphur-based SO_3/air sulphonation.

There are, however, major disadvantages using stabilised liquid SO_3:

- stabilised liquid SO_3 is a hazardous chemical (see section 3.3), which should be transported and handled with great care. Stabilised liquid SO_3 has the potential to present a major accident hazard;
- stabilised liquid SO_3 requires very precise temperature control and residues which contain noxious stabiliser can form. Residues after evaporation are difficult to handle and dispose of;
- liquid SO_3 raw material costs are higher compared with the costs of SO_3 raised from sulphur.

Figure 14 Stabilised liquid SO_3, dosing and vaporisation

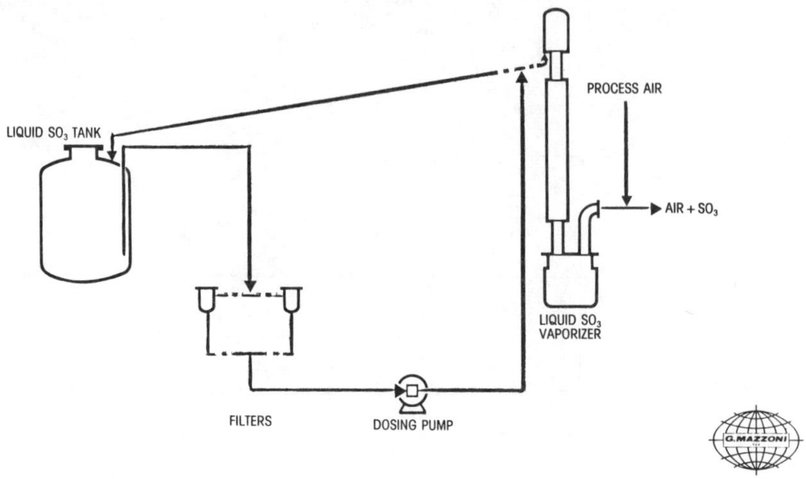

Figure 15 depicts a block diagram for SO_3/air sulphonation based on stabilised liquid SO_3.

Figure 15 Sulphonation-sulphation plant with stabilised liquid SO_3

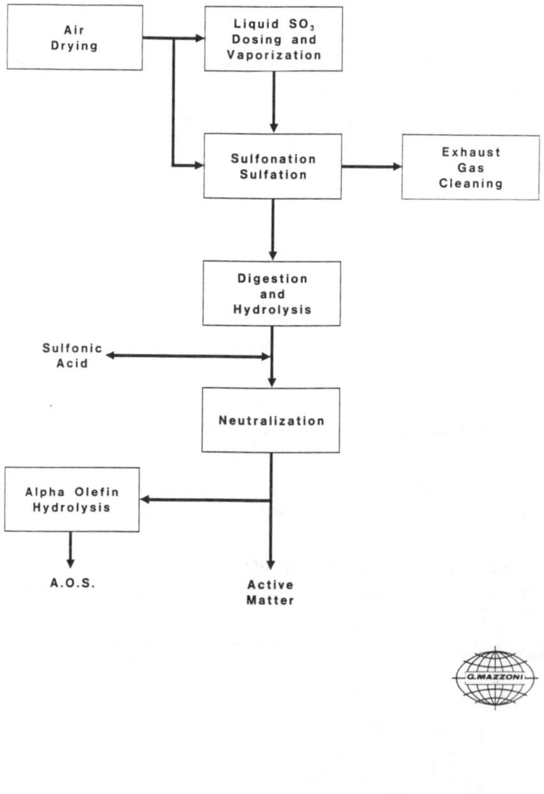

5.4 SO_3 generation from 65% oleum

The SO_3 is obtained by heating 65% oleum to approximately 250°C. The 65% oleum-containing stills can be externally heated at the base by gas jets and are well-lagged on top to prevent loss of heat. The optimum temperature of the still is 250°C. At this temperature the oleum loses its free SO_3 and the remaining sulphuric acid overflows from the stills into a store tank below. A seal of sulphuric acid is maintained in a "U" bend in the overflow line to avoid loss of suction from the stills. The purpose of the seals is to maintain a suction in the stills, i.e. avoid any air being drawn up from the sulphuric acid store tanks through the overflow line. The liberated SO_3 gas leaves the stills by means of a vertical reflux column from the centre of the upper surface of the still. The column acts as a condenser, removing any liquid SO_3 from the gas stream. The top of the SO_3 column is connected to an air mixer. In the mixer, the SO_3 stream meets an air stream introduced tangentially into the mixer in an opposite direction. The SO_3/air mixture is removed from the top of the mixer by a mild steel tube which feeds the mixture into the sulphonators.

The advantages of the 65% oleum route are the following:

- lower plant capital costs compared with sulphur-based SO_3/air sulphonation;
- reduced plant start-up/shut-down time spans and overall easier operation compared with sulphur-based SO_3/air sulphonation;

There are, however, major disadvantages using 65% oleum to raise SO_3 gas:

- 65% oleum is a hazardous chemical (see section 3.4.), which should be transported and handled with great care. 65% oleum has the potential to present a major accident hazard;
- the concentrated sulphuric spent acid has to be returned to the oleum supplier;
- 65% oleum raw material costs are higher than costs of SO_3 raised from sulphur.

5.5 Sulph(on)ation reactor systems

5.5.1 General chemical reaction engineering aspects related to sulphonation reactor systems

To understand the various sulphonation reactor designs, the following overall characteristics of SO_3/air sulphonation should be borne in mind.

- In chapter 4 the chemistry of sulphonating organic feedstocks was discussed. The reaction of organic feedstock (F) with two mole equivalents of SO_3 to produce a 2:1 species ($F + 2\ SO_3 \rightarrow F(SO_3)_2$) is usually so fast as to be almost instantaneous, and is highly exothermic.

- Not only the desired reaction leading to the organic acid takes place. Parallel and consecutive reactions may take place resulting in undesired products and colour bodies.

- The viscosity of the organic phase increases during the reaction, notably between the 70 and 100% conversion range.

From these general characteristics a number of criteria for sulphonation reactor types can be selected. The combination of an instantaneous reaction, highly exothermic, and a factor 50-100 increase in viscosity makes it clear that temperature control of the organic phase during the reaction is the main problem. Therefore the reaction rate has to be tempered by physical means to control the rate of reaction and therefore the organic phase temperature. By diluting the SO_3 in air, the SO_3 flux to the organic liquid interface will be reduced. In other words, the rate of reaction will be controlled by the transport rate of SO_3 in the gas-phase to the organic liquid phase; volumetric levels of 4-7% SO_3 (4% for say alcohol ethoxylates, 7% for high-quality linear alkylbenzenes) are therefore used. It is also clear that the heat of reaction has to be removed and that there is a need for extensive cooling area in the selected reactor system.

Conclusions:

(i) The bulk temperature of the organic phase and the liquid/gas interfacial temperature should be kept as close as possible to the optimum temperature to suppress the reaction rates of undesired reactions. Therefore, the rate of reaction should be controlled from the gas-phase by diluting the SO_3 with air (4-7 vol%) and the ratio of cooling surface area to reactor hold-up volume should be high.

Other reactor design criteria can be derived from the general sulphonation reaction chemistry.

(ii) To suppress parallel and consecutive undesired reactions the fresh SO_3/air feed should be prevented from coming into contact with partially converted product. Therefore, co-current flow of SO_3/air gas and the organic liquid phase is preferred, so that a decreasing SO_3 flux from the gas-phase to the organic liquid interface coincides with a decreasing presence of unreacted organic feedstock.

(iii) Back-mixing of the organic phase will result in contact at the gas/liquid interface of partially converted product with a relatively high SO_3 concentration in the gas-phase.

This may stimulate by-product formation. Therefore "plug-flow" of the organic liquid phase and the gas-phase in a co-current gas-liquid reactor would be preferred.

(iv) The time/temperature history must be optimal to maximise the production of the desired product. Extremely long reaction times (CSTR system) will result in undesired byproducts for certain feedstocks. Short residence time reactor systems are therefore preferred for the capability to handle all feedstocks.

(v) Macro back-mixing should be avoided but micromixing of the hot liquid interface with its direct surroundings will help to suppress hot spots and therefore by-product formation.

Summarising, the following reactor selection criteria can be listed.

Table 11 Summary of desired sulphonation reactor criteria

1. Diluted SO_3 in gas phase to temper the rate of reaction.
2. High ratio of cooling surface area to reactor hold-up volume.
3. Co-current flow of organic liquid and SO_3/air gas phase.
4. Plug flow for liquid and gas phases.
5. Short reaction-time reactor.
6. Good local mixing of the interface organic liquid with the bulk of liquid organic phase.
7. Mechanical simplicity of the reactor design and easy/reliable running characteristics would be an asset.

There are seven main sulphonation reactor systems used world-wide for SO_3/air sulphonation for which there is proven practical and documented experience: The Ballestra Sulphurex continuous stirred tank reactor (CSTR) cascade, the Ballestra Sulphurex F (a multitube falling-film reactor (MT-FFR)) the Chemithon falling-film reactor (FFR), the Chemithon Jet Impact Reactor, the MM falling-film reactor (FFR), the Mazzoni Sulpho film reactor (a multitube falling-film reactor) and the Japanese T-O FFR reactor system.

Table 12 SO_3/air sulphonation reactor systems used world wide

Equipment manufacturer	Reactor system name	Reactor type
Ballestra	Sulphurex Cascade	CSTR
Ballestra	Sulphurex F	MTFFR
Chemithon	Chemithon FFR	FFR
Chemithon	Chemithon Jet Impact reactor	Jet Reactor
Meccaniche Moderne	MM-FFR	FFR
Mazzoni	Sulpho film reactor	MT-FFR
Lion Corporation	T-O Reactor System	FFR

Legend:

CSTR: A Cascade of Cont's Stirred Tank Reactors
MT-FFR: Multi Tube Falling-Film Reactor
FFR: Falling-Film Reactor

The various types will be described in detail in the following paragraphs. At the end of chapter 5.5. the various reactor systems in use will be evaluated according to the 7 criteria of table 11.

5.5.2 The Ballestra Sulphurex CSTR System (fig. 16)

The Ballestra system consists of a number of stirred-tank reactors, arranged in a cascade sequence or train. Each reactor is equipped with a high-speed turbine impeller to disperse the gas and to mix and circulate the organic phase within the reactor. Cooling is facilitated by cooling coils located within the body of the reactor and a cooling jacket around the reactor.

The number of reactors within the cascade train is governed by the plant capacity but is normally from 4 to 6 units. The organic feed enters the first reactor and the liquid mass, under steady-state running conditions, overflows by gravity from one reactor to the next. SO_3/air is added to each reactor individually, the quantities added being controlled according to the duty of that particular unit. The conversion in any reactor can be controlled to provide optimum operation for each unit. Thus:

Table 13 Sulphurex CSTR sulphonation

Reactor in CSTR	1	2	3	4
Conversion per unit %	49	28	18	3
Cumulative conversion	49	77	95	98
Heat flow per reactor kJ/h $(x10^{-3})$	596	344	223	42
Air flow per reactor m^3/h	1300	750	480	80
Superficial velocity m/s	0.45	0.26	0.17	0.03
Cooling surface area m^2	15	12	10	3
Estimated overall heat-transfer coeff. $w/m^2\,^\circ C$	320	230	175	115

These data are based on an alkylbenzene feed of 8 kmol/h or 2000 kg/h (2800 kg 100% AD/h) and an SO_3 level in air of 6 vol%. Each stirred tank has a diameter of 1 m.

By careful selection of the reactor sizes and the SO_3 quantities added to each reactor in the series, full advantage can be taken of the benefits of maximising conversion in the initial reactors in the chain. Here the reaction mass still has a relatively low viscosity and thus full advantage can be taken of a high heat transfer coefficients. At 90% conversion, the heat transfer coefficient is only a quarter of that at 10% conversion, due to the increase of viscosity of organic phase with increasing conversion.

In the Ballestra CSTR system the flow of SO_3/air is reduced in each successive reactor i.e. achievement of conversion is biased to the early reactors in the train. However, the quantity of gas that can be fed to a reactor is limited to that at which flooding of the turbine impeller occurs which results in severe entrainment of the organic liquid with the reaction gas. The maximum superficial gas velocity is in the region of 1 m/s, based on the free cross-sectional area of the reactor. Because of this limitation, Ballestra CSTR plants tend to operate with relatively high SO_3-in-air concentrations (6 to 7% by volume).

The use of this higher SO_3 concentration is possible for linear alkylbenzenes, without significant quality penalties.

Back-mixing and short-circuiting of the liquid stream must inevitably occur to some extent in a cascade reactor system of the Ballestra type. Thus for a reactor train of 4 reactors where the mean residence time is 90 minutes, 15% of the material passes through in approx. 45 minutes, while 5% of the material has a residence time approaching 180 minutes.

<u>The Ballestra CSTR system, because of its back-mixing characteristics, long residence time and residence time distribution, can expose the organic material to by-product formation by side reactions.</u> (see 5.5.1.) <u>This can prove disadvantageous for sensitive materials such as primary alcohols, alcohol ethoxylates and alpha-olefins. Therefore, the Ballestra CSTR system is predominantly used for branched and linear alkylbenzenes and fatty acid methyl ester (FAME) sulphonation</u>.

Temperature hot spots do not occur in this system. The organic liquid film around the rising gas bubbles is extremely thin and continuously refreshed. The bulk organic liquid provides an instantaneous heat sink and the heat of reaction is efficiently removed by the cooling water in the cooling coil system within the reactor.

It can be calculated and it is found in practice that the SO_3 in the rising gas bubbles has been completely absorbed by the organic phase at the organic liquid-gas interface in the top of each stirred tank reactor, where the exhausted gas leaves the reactor.

Ballestra Sulphurex plants normally include an ageing vessel and a hydrator reactor (both of which are agitated and contain cooling water coils: see 5.5.10.) as part of the total stirred tank reactor train. Normally a stirred tank reactor for neutralisation is installed as the last stirred tank reactor in the total reactor train.

A small proportion of fresh organic feedstock can be bled into the last sulphonation reactor, before ageing, to adjust the conversion, thus avoiding the long response time caused by the extensive hold-up volume in the total system.

Figure 16 shows the Ballestra Sulphurex cascade system.

Figure 16 The Ballestra Sulphurex CSTR system

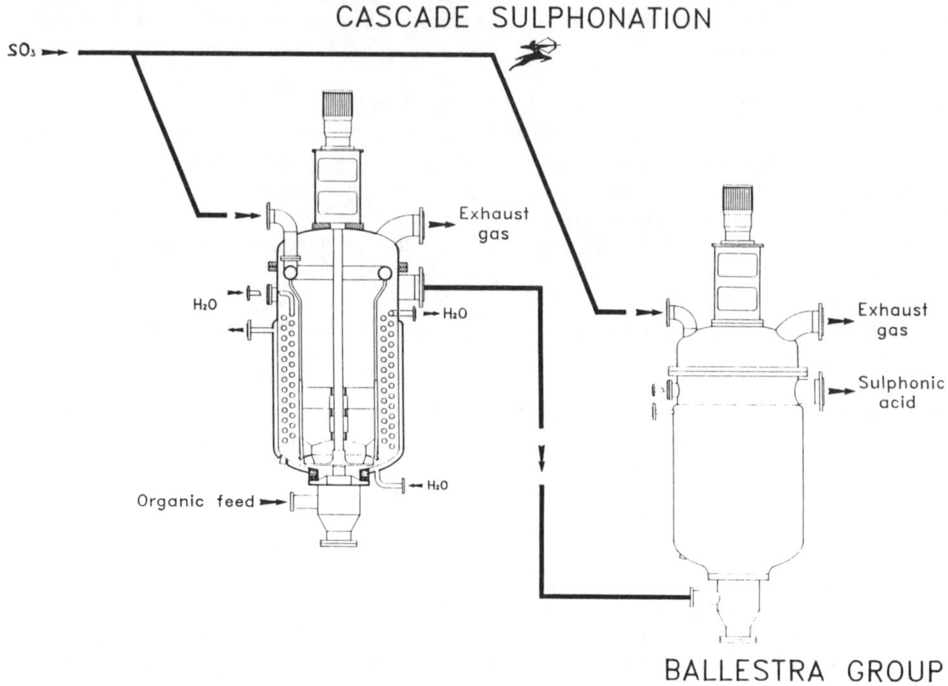

5.5.3 Chemical reaction engineering aspects related to sulph(on)ation in falling-film reactor

Assume a vertical tube with a thin organic film falling down along the wall from the top of the tube. Cocurrently with the organic phase, SO_3/air gas flows through the centre of the tube. Assume a gas velocity of 25 m/s. The initial SO_3 concentration in the gas is for example 4% and the outlet gas at the bottom is virtually exhausted of SO_3. The reaction rate is gas-phase controlled and it is assumed that any SO_3 molecule from the gas bulk phase arriving at the liquid organic phase will react immediately; i.e. the SO_3 concentration at the liquid interface, Ci SO_3, is zero. A simple mass balance over the reaction tube results now in the following equation.

ln Co/CL= 4 kg L/D.v in which

Co = SO_3 concentration in gas at entrance (4%)

CL = SO_3 concentration in gas at exit, assume
95% exhaustion CL = 0.05 x 4% = 0.2%

kg = mass transfer coeff. in the gas-phase (m/s) in the order of 7.5 x 10^{-2} m/s calculated from the Sherwood/Reynolds/ Schmidt correlation for turbulent flow through tubes. The diffusivity of SO_3 in air is estimated to be 1.5 x 10^{-5} m^2/s.

v = reaction gas velocity (m/s)

```
                   Co  = 4%vSO3                     Mass balance over
             ----|-------------|--|- x = 0          a pipe in the film
                 |             |  |                 reactor
                 |     v=25 m/s |
                 |        |    |  |                 Assumption: chemi-
                 |   Cx   ↓    |                    cal reaction at
             ----|-------------|--|- x              interface is much
                 |   Cx+dCx    |                    faster than mass
Ci=CSO3      ----|-------------|--|- x + dx         transfer of SO3
at interface|         D        |                    to the  interface
= 0             |<----------->|  |                  (gas-phase con-
                 |             |                    trolled absorption
                 |             |  |                 with an SO3 concen-
                 |             |                    tration at inter-
             ----|-------------|--|- x = 1          face Ci SO3 =0)
           CL=0.05 x 4% = 0.2%SO3
```

Mass balance

$$kg\ \pi D.dx.Cx = -\ \pi/4.D^2.v.dCx$$

$$dx = -\ D.v/4kg\ .\ dCx/Cx$$

$$\int_0^L [dx] = +D.v/4kg \int_{0.2\%}^{4\%} dCx/Cx$$

$$L = [D.v/4kg]\ \ln Co/CL \ ......(1)$$

L = Reactor tube length (m)
D = diameter of the tube (m)
v = gas velocity in the tube (m/s) in this example 25 m/s

From this simple equation the conversion of the organic can be calculated as a function of the reactorlength, x.
x = 0 corresponds with the top of the reactor and x = 6 m corresponds with the bottom of the reactor. The conversions are calculated for a kg value of 7.5 x 10^{-2} m/s and for a value of a kg' double this value of mass transfer as a result of the ripples on the liquid phase caused by the

high gas velocity.

The following table shows the results of this calculation.

Table 14 Conversion as a function of falling-film reactor length.

x from top (m)	conversion of organic (Co-Cx)/Co x 100%	
	1 x kg	2 x kg (= kg')
0.1	4.8	10.0
0.3	13.5	25.0
0.5	21.3	38.3
1.0	38.3	61.8
2.0	61.8	85.3
3.0	76.3	94.5
4.0	85.3	97.8
5.0	91.0	99.3
6.0	94.5	99.7

From this table it becomes clear that 20% of the conversion takes place in the first 0.30 m of the total reactor length, that 50% conversion is passed at 1 m from the top and that 75% is attained after 2.0 m from the top. In fig. 17 the conversion is plotted as a function of the reactor length. This curve is derived from a computer model (Ballestra) of the SO_3/air sulphonation in a falling-film reactor tube. This conversion characteristic has significant consequences for the temperature profile and the liquid interface temperature profile over the total reactor length. In fig. 18 the liquid interface temperature is indicated as a function of reactor height and in fig. 19 the "bulk" liquid temperature of the film is shown. Both curves indicate clearly a peak temperature about 0.5 m from the top of the reactor. This peak temperature is of significant importance for the product quality.

It is clear that an increase in liquid film thickness (i.e. a higher load of organic feedstock) will result in an increase in peak temperature. All equipment suppliers therefore have specified their organic feedstock close to 0.4 kg organic feedstock per mm circumference per hour. This is based on a feedstock of linear alkylbenzene.

The viscosity of the organic phase changes with conversion. In fig. 20 the viscosity increase is shown as a function of reactor length. The film thickness is strongly affected by the viscosity increase as can be seen in fig. 21. The initial film thickness is about 320 micron (0.32 mm). At the peak temperature the film thickness is about 500 micron and increases further to about 2000 micron (2 mm) at the reactor tube outlet.

The heat of reaction will be removed mainly by the cooling water around the reactor tube. Cooling is of great importance, notably at the top of the reactor. Normally two cooling zones are distinguished: one for the upper section and one for the lower. The optimal cooling water temperature has to be established for each kind of organic feedstock. The cooling water circulation rate is very high so that the heat resistance on the cooling water side is minimised and the temperature increase between incoming and outgoing cooling water is marginal (1 °C max.) to maintain a maximum driving force to remove the heat of the highly exothermic reaction.

It is not always possible to use very cold cooling water, due to the risk of freezing the organic feedstock or organic feedstock/organic acid mixture. Also the sharp increase of viscosity in the mixtures of organic feedstock/organic acid at low temperatures may cause an overall reduction of heat transfer when the cooling water is too cold. The upper and lower parts of a FFR require different cooling water flows and temperatures.

The gas phase absorbs some of the reaction heat in the top critical zone of the reactor. In fig. 22 the gas temperature is plotted against the reactor length.

Figure 17 FFR modelling: conversion as a function of reactor height

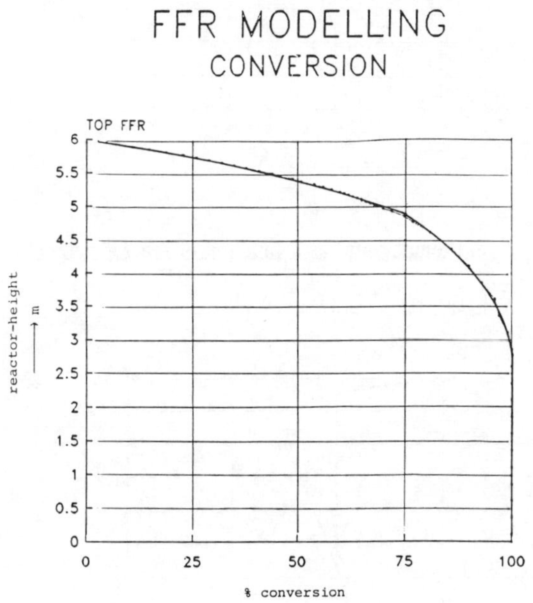

Figure 18 FFR modelling: interface temperature as a function of reactor height

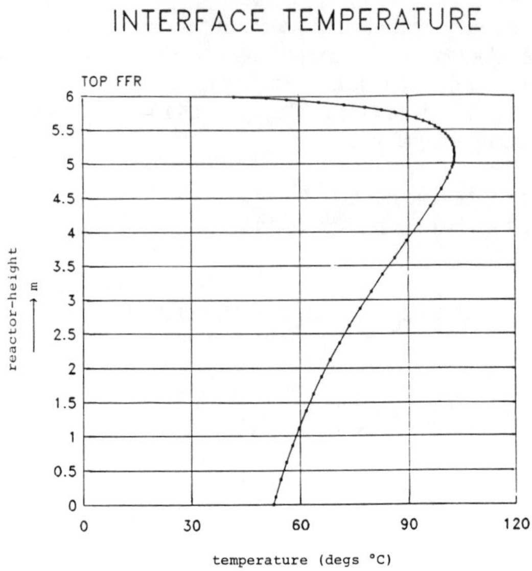

Figure 19 FFR modelling: liquid temperature as a function of reactor height

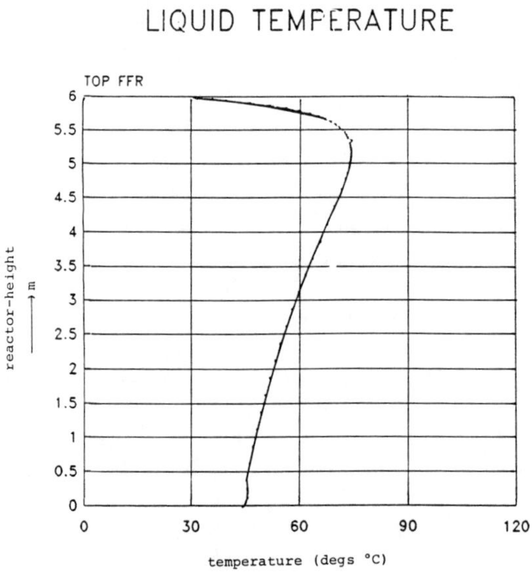

Figure 20 FFR modelling: liquid organic phase viscosity as a function of reactor height

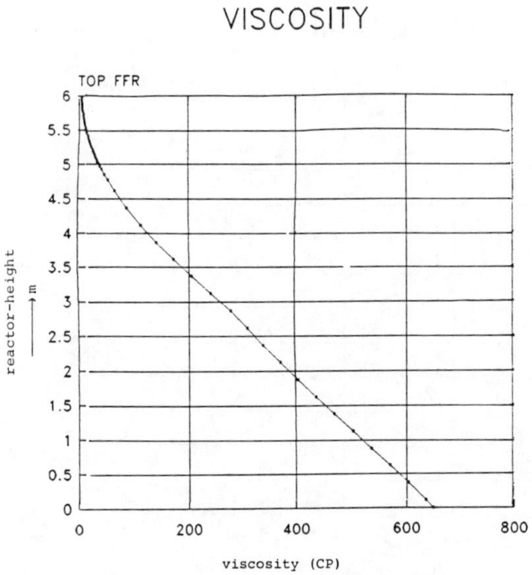

Figure 21 FFR modelling: film thickness as a function of reactor height

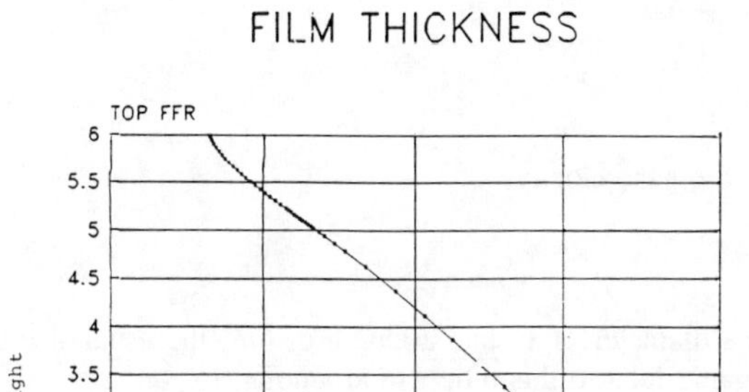

Figure 22 FFR modelling: reaction gas temperature as a function of reactor height

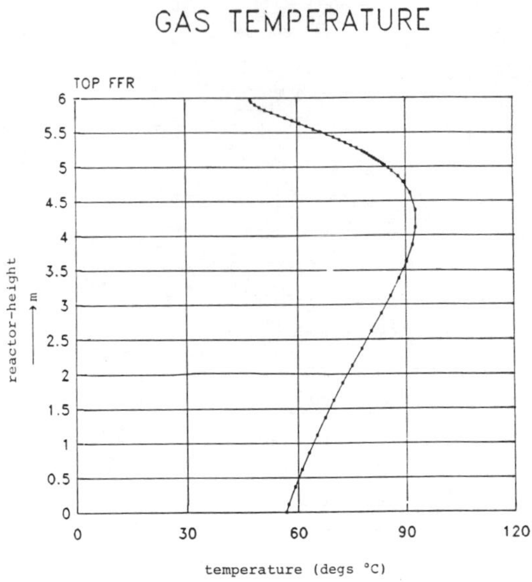

The massflow of SO_3 to the organic liquid interface is direct proportional to the product SO_3 concentration in gas and Φ_v the volumetric gas flow (m^3/s). This can be understood from the general mass transfer equation:

$$\Phi_m = kg\ A\Delta C$$

kg is almost proportional to the gas velocity;

$$kg \sim v^{0.8} \sim \Phi_v^{0.8}$$

Assume a reactor tube with a diameter of 1" and a length of 6m, the usual size in a multi-tube falling-film reactor. The organic feed to this tube would amount to

0.4 kg/mm circumference x π x 25.4 = 31.9 kg/h linear alkylbenzene with a MW of say 245. This would require 0.130 kmol SO_3/h corresponding to a volume of

0.130 x 22.4 x 323/273 x 1/1.3 = 2.65 m^3SO_3/h

at a temperature of 50°C and a pressure of 1.3 bar.

The total volumetric flow would become

2.65 x 100/5 = 53 m^3/h

at a level of 5% SO_3 and 37.9 m^3/h at a volumetric level of 7% SO_3 in air. The corresponding gas velocities would become 30.0 m/s and 21.5 m/s. Under both conditions the SO_3 flux to the interface is roughly the same ($\Phi_v.\Delta C$ is constant) and the same conversions would be obtained along the reactor length in both cases.

Let us now assume a tube reactor with a diameter of 2". The organic feed would double as the circumference doubled and therefore the total volumetric gasflow should be doubled, assuming constant SO_3 levels in the gas-phase.

$$\Phi_{v1} = \pi/4 \text{ x } d_1^2 \text{ x } v_1 \qquad \text{1" tube}$$

$$\Phi_{v2} = 2\Phi_{v1} = \pi/4 \text{ x } (2d_1)^2 \text{ x } v_2 \qquad \text{2" tube}$$

From above equations it becomes clear that $v_2 = 0.5\ v_1$.
It can be seen now that at the same SO_3 concentrations in air the kg_2 would become roughly a factor 2 smaller due to the fact that $v_2 = 0.5\ v_1$, i.e. at the same SO_3 concentration in gas the SO_3 flux to the interface would be halved in the 2" tube. Therefore double the reactor length would be required for the 2" tube to absorb all the SO_3. Alternatively, the gas velocity would have to be doubled to retain a length of 6 m, which is undesirable because of organic entrainment in the gas-phase leading to dark organic material by over-sulphonation. An alternative would be to double the SO_3 concentration from 5% to 10% and 7% to 14% respectively. High concentrations of SO_3 in air, however, are not feasible in sulphur-based SO_3/air gas raising plants.

By reducing the 1" tube by a factor two the opposite effect would be obtained: the tube length could be reduced to 3 m or the SO_3 concentration could be reduced to 2.5% or 3.5% or the gas velocity halved whilst maintaining a length of 6 m. However, many reactor tubes of 1/2" would be required for a reactor of practical throughput rate.

The above example illustrates that changing falling-film reactor design (diameter, length) has consequences in terms of required volumetric flow and gas concentration, assuming a max allowable organic feed per mm circumference per hour. This latter parameter can not be manipulated upwards due to heat transfer limitations.

Film reactors are short residence time reactors. The mean residence time of the organic film in this type of sulphonation reactors is in the order of 30 seconds, which compares with the CSTR design with a mean residence time of 60-90 minutes!
Many attempts have been made to model in mathematical terms the SO_3/air sulphonation of organic feedstocks in falling-film reactors. A few examples are described below.

Johnson and Crynes (1974) published the results of a mathematical model in which organic bulk liquid temperatures and the gas bulk temperature are predicted along the 3 metres reactor length. For this mathematical model the following assumptions were made:

(i) no entrainment of liquid droplets into the gas core;
(ii) no entrainment of gas bubbles into the liquid film;
(iii) turbulent flow in both phases (organic phase and gas phase;
(v) reaction rate completely gas-phase controlled.

The model predicts peak temperatures about 0.5 metres from the top of the reactor for the liquid/gas interface, the bulk liquid and, to a lesser extent, for the gas-phase. A conversion of about 90% is attained at 1 metre from the top. The organic liquid/gas interfacial temperature was obtained by assuming a linear temperature profile in the film, which conflicts with the assumption of turbulent flow of the organic phase.

Davis, Ouwerkerk and Venkatesh developed a mathematical model to predict the conversion and temperature distribution in the reactor as a function of the gas and liquid flow rates, physical properties, the feed composition of the reactive gas and carrier gas and other parameters of the system. Transverse and axial temperature profiles are calculated for the laminar flow of the liquid phase with co-current flow of a turbulent gas to establish the peak temperatures in the reactor as a function of the numerous parameters of the system. Also in this model, the reaction rate in the liquid film is considered to be controlled by the rate of transport of reactive gas from the turbulent gas mixture to the gas - liquid interface. The predicted reactor characteristics are shown to agree with large-scale reactor performance. For the calculations of the mass transfer coefficient in the gas phase, kg, Davis et al. used the same correlation as Johnson and Crynes, but multiplied the calculated values arbitrarily by a factor 2 to include the effect of ripples on the organic liquid film caused by the high SO_3/air velocities in the core of the reactor.

A recent model (1988) was published by J. Guttierrez Gonzalez et al. According to the authors, although the liquid flow is laminar, due to the high Schmidt number in the liquid phase, eddy mass transfer can be significant and eddy diffusion cannot be disregarded with respect to molecular diffusion. Eddy thermal diffusion in the liquid phase is much smaller than thermal diffusion, so that it is not introduced in the microscopic heat balance. The Spanish authors needed to validate their results on practical data. The pressure drop over the reactor, heat- and mass transfer were fitted.

Beenackers, Fok, Van der Velde and Nanninga (1986 - 1988) of the Groningen University in the Netherlands developed together with Ballestra in Milano a model similar to the approach of Guttierez Gonzalez and co-workers. Total gas phase limitation throughout the falling-film tube reactor is assumed and the reaction takes place at the interface only. The concentration profile of the organic raw material in radial direction was not considered. Again most of the computer runs were carried out for linear alkylbenzene (LAB). Like in the previous models, the variation in the molefraction SO_3 most strongly influences the peak liquid temperatures. Beenackers et al. also need to validate their results based on practical data. This mathematical model has not been published yet, but results of the calculations are depicted in this section (figure 17 through figure 22).

5.5.4 The Ballestra Sulphurex F system (multi-tube falling-film reactor MTFFR) (fig. 23 and fig. 24)

The complete reactor resembles the design of a shell-and-tube heat exchanger (see fig. 23). Parallel stainless steel 1" tubes are arranged vertically within a shell where the cooling of the reactor tubes takes place. The length of the reactor tubes is 6 m. There is a top cooling section and a bottom cooling section. Cooling water temperature can be controlled at the desired value. The organic feedstock is metered via a common liquid feeding chamber, maintained completely full, into a plurality of identical annular slots so as to form a film of the organic liquid on the internal surface of each tube.

The annular distribution slots for each tube are calibrated by the supplier before the tubes are fitted in the reactor. The width of the slits is selected in such a way that all organic feedstocks can be used without any re-setting of the slit. Reactor tube calibration is checked prior to plant start-up. During annual inspection the width can be re-set. The SO_3/air gas flows into the tubes above the annular slots from a common gas-feeding chamber on top of the tubes, the feeding pressure of the SO_3/air gas being in the range 0.2-0.3 bar gauge.

Figure 23 Ballestra, multi-tube falling-film reactor (MT-FFR)

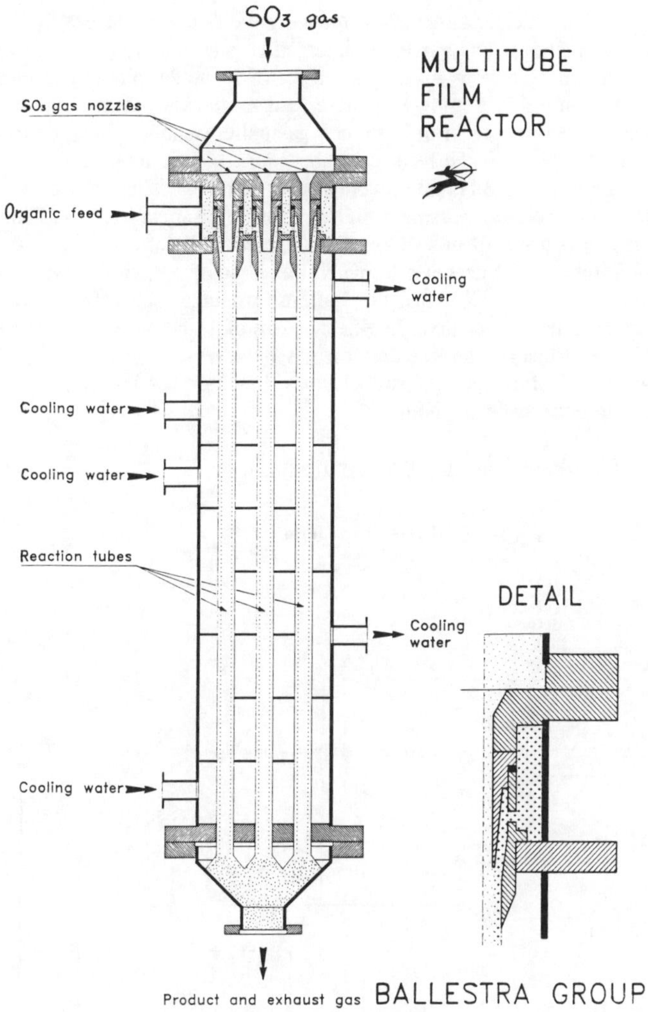

There are no "nozzles" for an even SO_3/air distribution over all pipes. The pressure drop over the 6 m pipe can be calculated with the well-known Fanning equation for turbulent flow. For an SO_3/air velocity of 30 m/s in an empty reactor pipe a value of 0.025 bar is found. In experiments with air and organic film (without any sulphonation reaction) a pressure drop of 0.1 bar was measured at an air velocity of 30 m/s. The interaction of the air with liquid film ripples and waves causes an increase of about a factor 4 in the pressure drop compared with the empty tube. Under real sulphonation conditions the pressure drop measured becomes even greater at 0.25 bar, due to the sharp increase of viscosity at high conversions.

With rising viscosity, the film thickness increases, reducing the free area for gas flow and increasing the gas film interaction. This gas liquid interaction is an important effect of the so-called "self-compensating" phenomenon. Assume a reactor of 3 parallel pipes. The total gas pressure drop over all these pipes is $\Delta P = P_1 - P_2$, where P_1 is the pressure in the common gas feeding chamber and P_2 is the pressure in the common gas outlet chamber. Imagine that, for example, tube 1 receives 5% less organic feed than tube 2 and 3. Assuming initially the same flow of SO_3/air in each tube, oversulphonation would occur in tube 1. This, however, would result in a sharp increase in viscosity, causing a higher pressure drop over tube 1. This is not possible because ΔP is constant for all tubes (common chamber at both ends of reactor tubes), therefore the gas flow to tube 1 will decrease to match the value of ΔP. In this way a self compensation effect will occur. It is clear that this Ballestra patented "self-compensating" effect will disappear for tube diameters wider than the selected size of 1" because the increasing film thickness will have a lesser effect on the free area for gas flow.
A 3 ton AD/h plant would require about 70 parallel tubes, based on 0.4 kg organic alkylbenzenes per mm circumference per hour.

Figure 24 Ballestra MT-FFR sulphonation plant arrangement

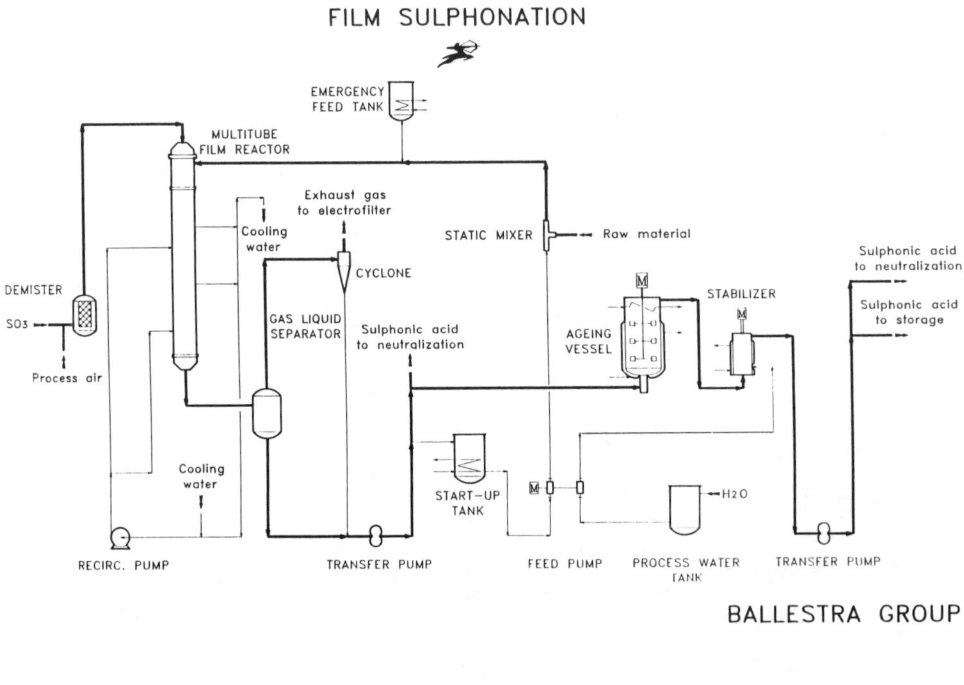

5.5.5 The Mazzoni Sulpho film reactor (multi-tube falling-film reactor, fig. 25)

The Mazzoni MT-FFR reactor system resembles the Ballestra design: parallel stainless steel 1" tubes are vertically arranged within a shell where the cooling of the reactor tubes takes place. The length of the reactor tubes is again 6m. In the Mazzoni system there are also two cooling sections: a top section and a bottom section. However, the difference with the Ballestra design is that each tube has its own cooling mantle. In the Mazzoni design, the SO_3/air reaction gas enters the top chamber of the reactor tube nest and is distributed through a venturi-type restriction to each pipe in the multi-tube reactor. The venturi in the top of each tube has a diameter of 13.88 mm and a length of 415 mm, the total pressure drop over this restriction being 0.4 bar according to Mazzoni design engineers. The pressure drop over the film reactor itself under running conditions is about 0.25 bar. In other words each pipe will receive the same amount of SO_3/air, as the main pressure drop component of 0.4 bar is a stationary, fixed restriction identical for each pipe. The organic feed enters in the middle chamber and distribution takes place through stationary nozzles made of agate stone, which have virtually unlimited lifetime.

A special characteristic of the Mazzoni MTFFR is the so called "equalizing air". Equalizing air enters the lower chamber at the top of the reactor. This dry process air serves two functions:

(i) the equalizing air presses the organic film against the wall.

(ii) the equalizing air forms a cushion between the SO_3/air and the organic phase, resulting in a less steep temperature peak in the top part of the film reactor and a more balanced reaction rate over the total length of the reactor.

Any dry air taken from the process to be used as equalizing air has to be compensated by the total amount of dry air i.e. the SO_3 level in the remaining air will increase proportional to the amount of SO_3-free equalizing air. The benefits of equalizing air have never been fully demonstrated in practice by model experiments.

The exhausted gas and organic acid mixture which leave the bottom of the reactor flow into two high-efficiency liquid-gas separators in series. The exhaust gas then flows to the exhaust gas treatment plant. The organic acid is transferred to the next stage, ageing and stabilisation of alkylbenzene sulphonic acid, or straight to the neutralisation unit for alcohols, alcohol ethoxylates and alpha-olefin feedstocks.

Figure 25 Mazzoni Multi-tube falling-film reactor (MT-FFR)

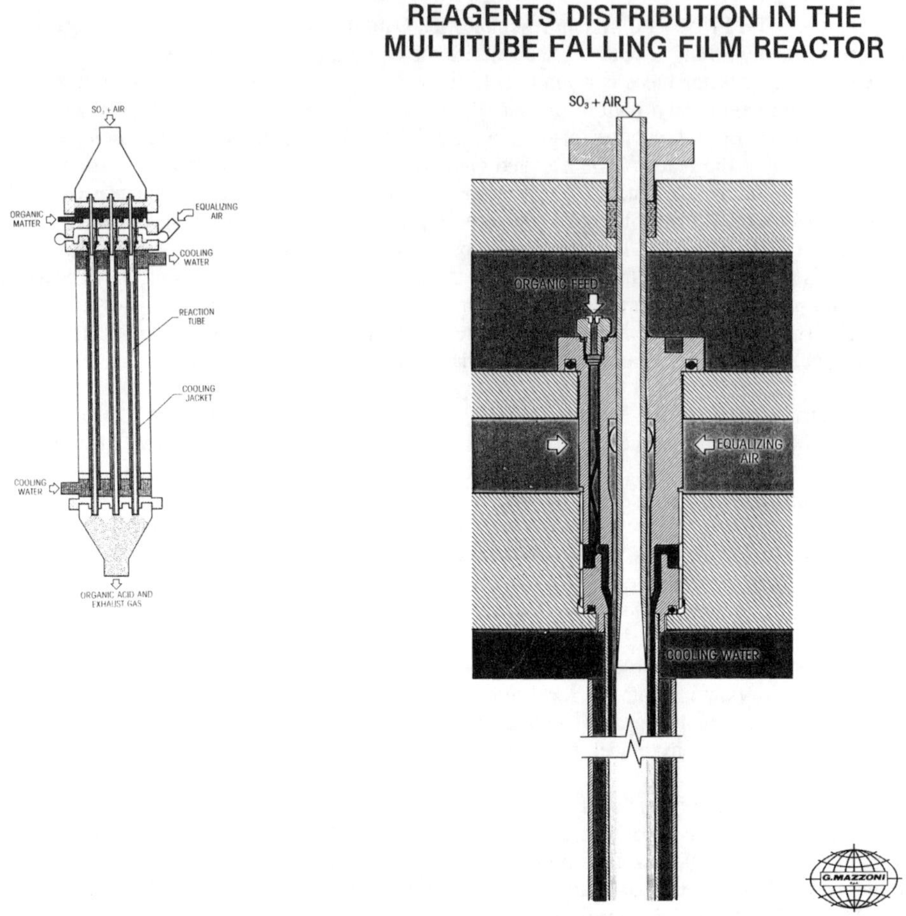

5.5.6 The MM falling-film reactor (fig. 26)

The Meccaniche Moderne (MM) reactor was the first falling film sulphonation reactor on the open market in the early seventies. This type of reactor is a direct offspring of the FFR worldwide patented by the Allied Chemical Corporation (USA) some years earlier.
The smaller (80 kg/h) MM reactor consists of a 6m long, 35mm diameter tube, practically just one tube of the later developed multi-tube FFR reactors. In the MM design the capacity increase of the reactor is not realised by increasing the number of parallel tubes but by enlarging the diameter of one tube, i.e. the MM FFR reactor is a so-called "Mono Tube FFR Reactor", consisting of two concentric 6m long cylinders. The organic feedstock flows down on the outside of the inner-tube and the inside of the outer-tube. In the gap between the two concentric tubes the SO_3/air flows from the top co-currently with the liquid films. The heat of the sulphonation reaction is removed via cooling-water flowing through the inside part of the inner-tube and the outside part of the outer-tube.

Figure 26 Meccaniche Moderne concentric tube FFR

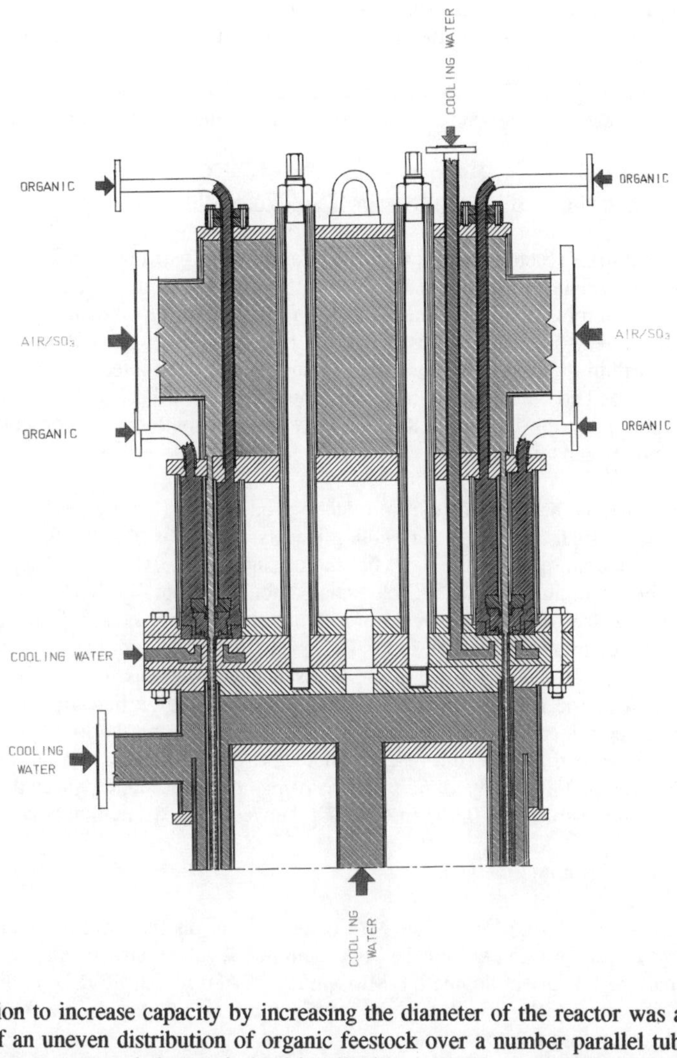

The solution to increase capacity by increasing the diameter of the reactor was adopted to avoid the risk of an uneven distribution of organic feestock over a number parallel tubes. The MM FFR reactor recommended organic load is still, like in the original Allied FFR reactor, about 0.4 kg Alkylate per mm reactor circumference per hour and also the length is still like in the original, that is 6m. It is no supprise that the required reactor length is exactly the same as for the Ballestra and Mazzoni MT-FFR. For the same air velocity and SO_3 concentration in the gas phase, the conversion of the same amount of organic per mm circumference per hour will result in the same required reactor length to exhaust the SO_3 from the gas phase.
For a 2500 kg AD/h plant (LABS) the MM FFR reactor has an inner tube of 665mm diameter and an outer-tube of 700mm, a 4000 kg AD/h plant has an inner-tube of 993mm diameter and an outer-tube of 1028mm. The annular gap is always the same: 16.5mm.

With the MM "Mono Tube FFR reactor" scaling-up takes place in an ever increasing reactor diameter that leads to an increase of the ratio Construction costs/Production capacity. Also the uniform distribution of the organic feedstock over the outer-inside and inner-outside concentric tubes with the increase of the reactor diameter require more and more precise tool machining of the distribution rings.
The maximum size single MM reactor in the market has a capacity of 4 tonnnes AD/h.
At the bottom of the reactor the organic acid is separated from the exit gasstream in a gas-liquid separator.

5.5.7 The Chemithon Falling-film reactor (Fig. 27 and 28)

The Chemithon falling-film reactor consists of a short falling-film part of about 2 m length and a loop reactor with heat exchanger and circulation pump. One could say it is the combination of a very short residence time plug-flow type reactor and a perfect mixer type reactor, in series. The falling-film part of the reactor system consists, like the MM reactor, of two cylinders arranged concentrically, one within the other, in which the organic feed is distributed over both pipe walls (inner wall of outer pipe and outer wall of inner pipe) via liquid distribution slits. Precision calibrated, replaceable metering flanges control the flow and distribution of the organic over the inner and outer annular reaction surfaces.

It is essential to remove the heat of reaction in order to produce high quality product. Both the inner and outer reaction surfaces have water cooling jackets. At the base of the film section the organic acid has a temperature of about 75°C, but the organic acid leaving the film part is immediately quenched with an excess of recycled cooled acid of about 40-50°C. The organic liquid is separated from the exhaust gas stream in a liquid/air separator tank, which can also act as a holding tank in the circulation/quench loop. The quench loop incorporates a circulation pump and an in-line water-cooled shell-and-tube heat exchanger and is used as a heat sink and rapid quench for the material leaving the film reaction section. A high-efficiency (99.8%) cyclone separates the entrained organic acid from the gas stream leaving the loop/quench part of the reactor. The organic load for the film reactor amounts to about 0.40 kg/mm circumference per hour. For a 2.5 ton AD/h (alkylbenzene sulphonate) plant, the outside cylinder diameter is 0.9144 m and the inside diameter is 0.900 m. The gap between the two cylinders is therefore

$$(914.4 - 900.0) / 2 = 7.2 \text{ mm}.$$

This compares with a gap of 16.5 mm on the MM reactor. It means that with the same organic load of 0.4 kg/mm circumference per hour and the same SO_3 levels in air, the gas velocity in the Chemithon film part will be about double the MM values, say 80 m/s at 3% SO_3 in air and 48 m/s at 5% SO_3 in air, i.e. the SO_3 flux to the interface will be roughly double the amount as compared with the other film reactors discussed.
(SO_3 flux to interface is directly proportional to $v^{0.8}$ and SO_3 concentration in the gas-phase, see 5.5.3.)

Conversion is virtually complete at the bottom of the film part with the remaining conversion taking place in the quench loop under mild temperature conditions.

These higher gas velocities also thin the organic film and further increase the heat transfer rate through the reactor wall. Care has been taken in the design of the upper portion of the reactor to ensure that the organic film does not strip off the wall of the reactor. This is important because stripping off the film (misting) leads to loss of mole ratio control in the reactor when the mist droplets react with the SO_3 gas.

The estimated mean residence time of the organic material within the falling-film part is around 10 to 15 seconds whilst the mean residence time in the recirculation loop can be adjusted from a few minutes up to 15 minutes by the level of organic acid material in the gas-liquid separation vessel and the cooling efficiency (hold-up in heat exchanger).

Figure 27 Chemithon concentric tube FFR with quench loop in series

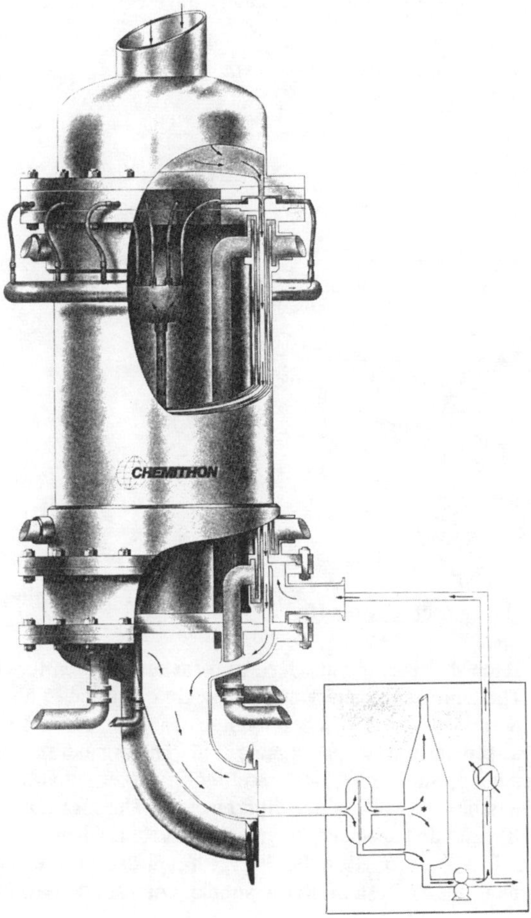

Figure 28 Chemithon details of concentric tube FFR

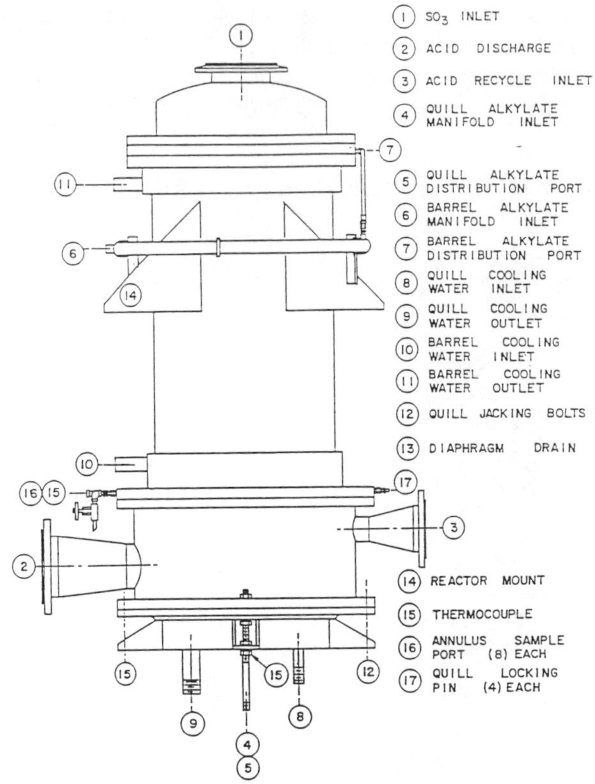

5.5.8 The Chemithon Jet Impact Reactor (fig. 29)

The Chemithon Jet Impact Reactor atomizes organic feed to form surface area for the SO_3/gas to react with the organic phase. The atomised organic mist is quickly quenched in a cooled recycle stream equivalent to the FFR described in 5.5.7. The product stream passes from the reactor to a cyclone separator where the gas separates form the organic acid. The organic mist droplets will reach high temperatures before quenching takes place, resulting in darker coloured sulphated products from this reactor type compared with falling-film reactors. The Jet Impact Reactor is therefore normally used for industrial and specialty chemical production when product colour requirements are not critical. It is also accepted in the detergent industry for the sulphonation of alkylbenzene feedstocks. The total reactor system is of a simple, compact design.

Figure 29 The Chemithon Jet-Impact sulphonation reactor system

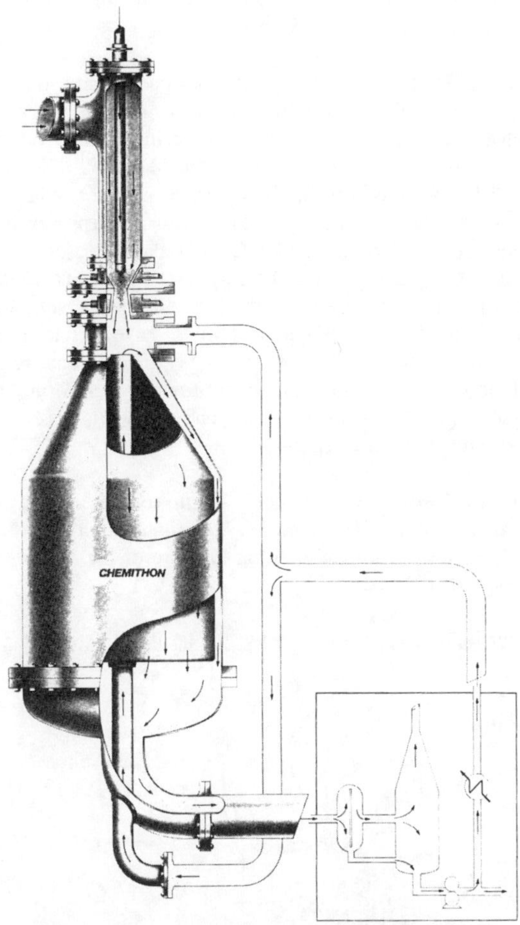

5.5.9 The T-O sulphonation reactor system (figure 30)

Examination of the technical and patent literature shows that the development of the T-O reactor has gone hand-in-hand with the establishment of an understanding of the process and mechanism of sulphonation of alpha-olefins. The reaction is very exothermic (ΔH = -210 kJ/gmole, approx. 30% more than that of alkylbenzenes) and virtually instantaneous. Lion Corporation were very active in the development of thin falling-film technology in the seventies, adding further refinements, which has made the large-scale production of AOS very attractive. The T-O reactor is the result of their R&D in this field and has been developed primarily for production of AOS.

The reactor is of the thin annular falling-film type, similar to the Chemithon reactor. It has a short length of about 2 m and a diameter of typically 0.3 to 1.0 m, depending on the capacity of the reactor. The width of the annulus through which the liquid film flows is about 5.5 to 30 mm. At the base of the film reactor, the reaction mixture is immediately quenched with an excess of recycled cooled acid of about 40 - 50°C, similar to the Chemithon set-up. The organic liquid is separated from the exhaust gas stream in a liquid/air separation tank. The quench loop incorporates a circulation pump and a heat exchanger.

As in other FFRs, the SO_3/air is introduced to the reactor concurrently in the downward direction. However, the method of distribution of SO_3/air is different. There is a special arrangement at the top of the reactor which separates the main SO_3/air from the organic film surface by a stream of air or air containing a small amount of SO_3 (0.02 - 1 vol.%). In this way, a resistance is set up for diffusion of SO_3 to the film surface, which results in a decrease in the rate of conversion at the initial stages in the top of the reactor. Sharp peak temperatures in this region are therefore somewhat smoothed out and T-O Lion claim an isothermal reaction. The concentration of SO_3 in the gas stream entering the core of the reactor is higher than normal (it can be as high as 16 vol.%!), but if it were to be diluted with the secondary air (whose volumetric flow rate is 2 - 7 times that of SO_3/air), it would have a concentration similar to that used in other falling-film reactors (i.e. up to, say, 5 vol.%). The organic feed is distributed on the wall by a specially designed slit type distributor. It provides a uniform distribution of organic liquid across the perimeter at very low pressure drop. The exhaust gases from the reactor can be partially recycled to the reactor after demisting.

The T-O FFR is an interesting design, but it is not a simple reactor in terms of mechanical construction and maintenance. For sulphur-based SO_3/air sulphonation, there are restrictions in the maximum level of SO_3 in the gas stream entering the core of the reactor. The T-O FFR has found restricted application outside Japan.

Figure 30 T-O reactor details of concentric tube reactor head with secondary air

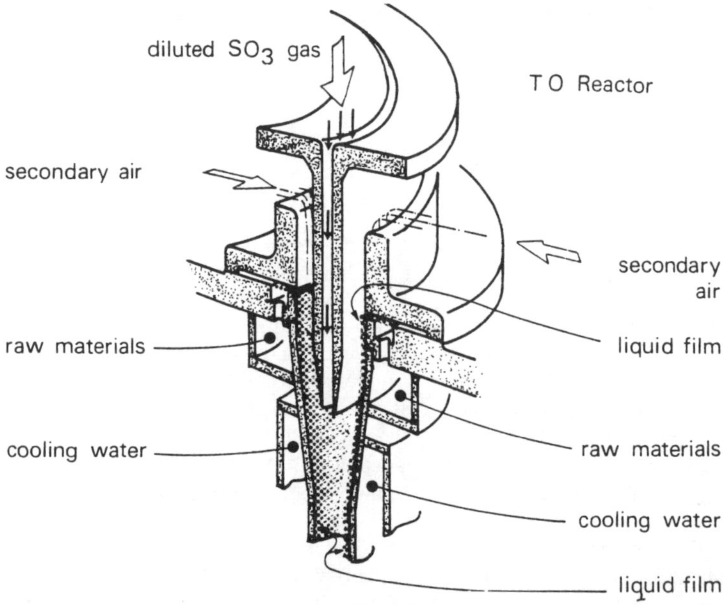

5.5.10 Ageing (Digesting) and hydrolysing (stabilising) equipment for alkylbenzene sulphonic acid (see fig. 31 and 32)

When sulphonating alkylbenzenes the sulphonic acid is submitted to ageing and hydration to convert the formed anhydrides (see 4.2.1.) to alkylbenzene sulphonic acid.

During ageing the pyro-sulphonic acid reacts with non-converted alkylbenzene to form two moles of sulphonic acid and reaction heat! The ageing tank should therefore be fitted with an agitator and cooling coils so that this post-reaction occurs under isothermal conditions! The ageing time may vary between different alkylbenzene feedstocks and therefore a design where the optimum digesting time can be chosen is preferred. The required digestion time is mostly in the order of 20 minutes at 45 to 50°C. For a very low H_2SO_4 content (0.5% max.) a longer residence time (hours) may be required.

Figure 31 Ageing vessel

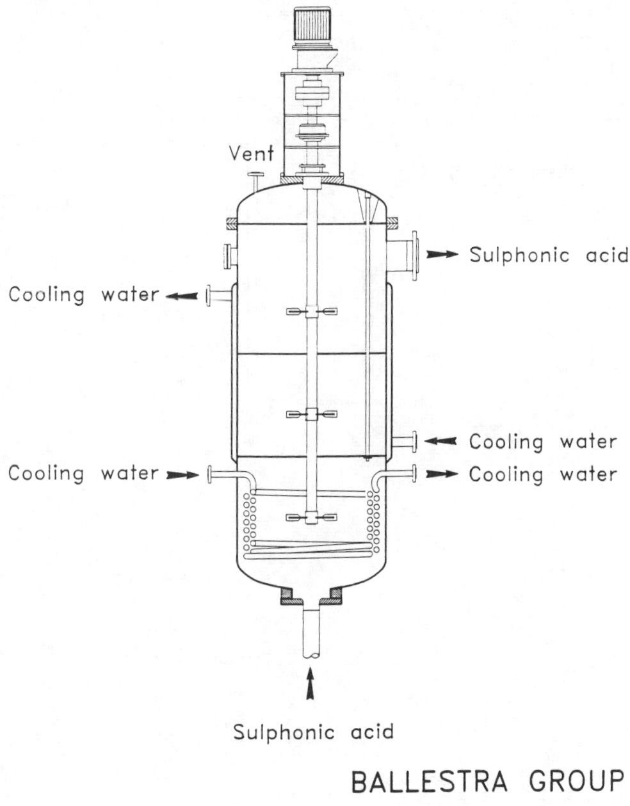

In the hydrolysing (stabilisation) step the organic anhydrides formed during sulphonation are converted with a small amount of water (about 0.3 to 1% on sulphonic acid) to alkylbenzene sulphonic acid and traces of loosely chemical bound SO_3 with water to form a small amount of sulphuric acid and reaction heat. The stabilisation tank should therefore be fitted with an agitator and cooling coils so that this post-reaction occurs under isothermal conditions.

Figure 32 Stabilising unit

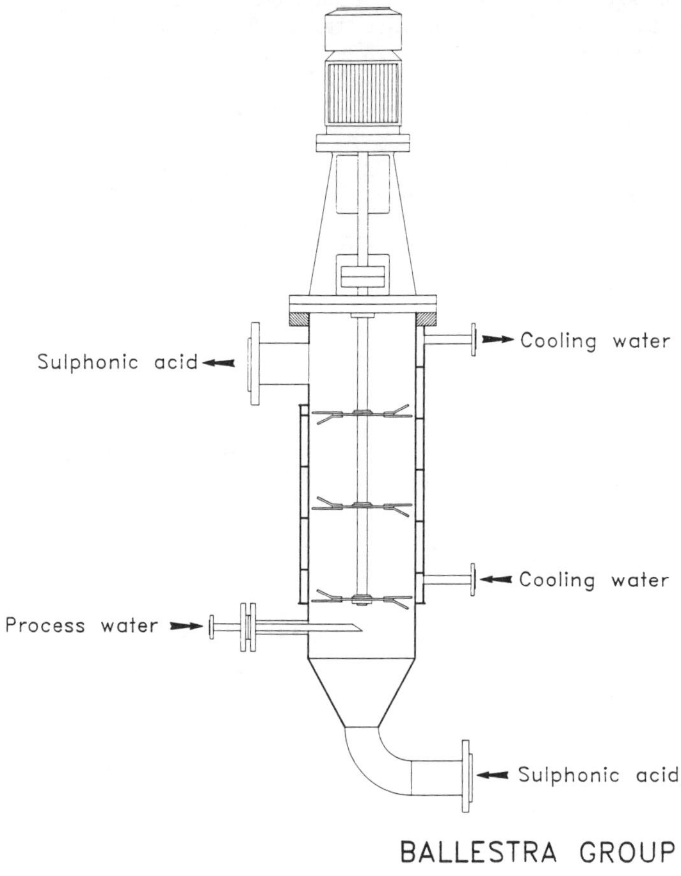

5.5.11 Desired criteria for sulphonation reactor design versus sulphonation reactor systems from suppliers (see table 15)

The design criteria for sulphonation reactors (5.5.1.) are compared against available designs in table 15.

1. Dilution of the SO_3 in the gas phase to temper the reaction rate is achieved in all suppliers' systems.

2. Co-current flow of the organic feedstock and the SO_3/air is realised in all falling-film reactor types and the Chemithon Jet Impact Reactor.

3. Plug-flow of the liquid phase is certainly not realised in the well-mixed liquid phase in the Ballestra CSTR system, although one could argue that overall plug-flow over the total of 4 reactors in series is approached. However, back-mixing in each individual reactor takes place, resulting in a high SO_3 flux to already-converted organic feedstock. "Close to" is perhaps a little bit optimistic for the falling-film reactor. The velocity profile of a laminar falling-film is a half-parabola with the maximum velocity at the film/air interface and a zero velocity at the cooling wall. However, due to the gas/liquid interaction, the organic film will be rippled and local mixing of interface and "bulk" may occur. This will contribute to plug-flow characteristics.

4. The pure falling-film reactors will have the shortest residence time, order of magnitude 30 seconds. The Chemithon FFR and Jet Impact Reactor, both with quench loop, could have an organic liquid residence time of about 2 to 5 minutes, depending on the liquid level in the de-gassing tank. Finally the Ballestra CSTR system shows a total residence time for the organic liquid of the order of 60 minutes.

5. A high cooling surface area/reactor hold-up ratio is established in all sulphonation reactor systems. However, a peak temperature in the organic phase in the top region of all film reactors (and the Jet Impact Reactor) can not be avoided.

6. Back-mixing of the organic phase would be beneficial for temperature equalization and heat transport. It would also help to bring fresh feedstock to the reaction front and transport converted material from the reaction front. In a CSTR system where SO_3/air gas bubbles rise to the organic phase this will happen. The liquid film around the rising gas bubbles is not only an order of magnitude thinner than the liquid film on the metalfixed wall of falling-film reactors but also gas bubbles are changing their liquid "coats" very frequently. In all falling-film reactors local mixing may happen due to the gas/liquid interaction causing surface ripples.

7. The Ballestra Sulphurex CSTR system is relatively simple in its mechanical construction. The same can be said about the Chemithon Jet Impact Reactor. Film reactors are mechanically more sophisticated constructions, notably the organic liquid distribution systems. The delicate FFRs can not always be recommended for countries where the level of technology and maintenance is low.

8. All systems are reliable and easy to run. However, falling-film reactors need more attention and precise procedures of operation. Uneven liquid distribution leads to poor

quality products and carbonisation of feedstock in the top part of the reactor. Cleaning procedures as outlined by the equipment suppliers should be strictly followed. Generally, reactor cleaning with water should not be done because severe corrosion of the vital distribution system may result.

During production stops, organic feedstock should be circulated over the film reactor to absorb residual SO_3 and to clean the reactor in the time that production is ceased.

Table 15 Desired criteria versus sulphonation reactor systems in practice

	Ball. CSTR	Ball. MT-FFR	Chem. FFR	Chem. Impact Jet	Mazz. MT-FFR	MM FFR
1. Dilution SO_3 in gas phase	Yes	Yes	Yes	Yes	Yes	Yes
2. Co-current flow SO_3 - organic feedstock	No	Yes	Yes	Yes	Yes	Yes
3. "Plug flow" for liquid phase	No	Close to	Close to	Close to	Close to	Close to
4. Short reaction time	No	Yes	Yes	Yes	Yes	Yes
5. High cooling ratio	Yes	Yes	Yes	1st part no - 2nd part yes	Yes	Yes
6. Good local mixing	Yes	Moderate	Moderate	1st part no - 2nd part yes	Moderate	Moderate
7. Mechanical simplicity	Yes	No	No	Yes	No	No
8. Easy reliable running	Yes	Yes	Yes	Yes	Yes	Yes

5.5.12 Conclusions: sulphonation reactor systems

There is hardly any difference in product quality between the various types of falling-film reactors from the various equipment suppliers. This is not so surprising, because:

(i) the recommended organic feed rate per mm circumference per hour is the same for all reactor types and the air velocity and SO_3 level in air are similar, i.e. the film thickness (heat transfer) and SO_3 flux to the interface (reaction rate) are roughly the same for all types of falling-film reactors. The exception is the Chemithon FFR-quench loop reactor where the SO_3 flux from the gas phase to the liquid organic phase is a factor 2 greater than for all other FFR type reactors; realising a quench loop to cool the FFR product;

(ii) even if small differences in product quality were to exist, the influence of differences in organic feedstock quality or variation in the quality of the total plant operation (sulphur quality, process air quality, SO_3/air filtering, operational temperatures, process control quality and instrumentation, etc.) would overshadow the differences in film reactor design.

The Ballestra Sulphurex CSTR and the Chemithon Jet Impact Reactor are usually only used for simple feedstocks like alkylbenzene.

5.6 Neutralisation equipment

5.6.1 Chemical Engineering aspects related to neutralisation reactor systems

To understand the various neutralisation reactor designs, the following overall characteristics of the organic acid neutralisation should be borne in mind.

The reaction between the organic acid and the neutralising agent, for example a caustic soda solution, is instantaneous, highly exothermic (see chapter 4.3) and the resulting paste of 40-75% active detergent (depending on type of feedstock) is highly viscous.

Paste viscosity can be expressed as follows:

$$\eta = Kj^{n-1}$$

where

η = the apparent viscosity (Ns/m^2)
K = consistency, that is the apparent viscosity at a shear rate of 1 s^{-1}. (Ns^n/m^2)
j = is the shear rate (s^{-1})
n = power law index, $n<1$, for pastes in the range 0.2-0.3

In other words, active detergent pastes are non-Newtonian so-called pseudoplastic liquors, with an apparent viscosity η, varying with shear rate. The values for K and n depend on AD concentration in the paste, temperature and type of active. K and n can be measured with a Haake Rotovisco meter.

In table 16 approximate values of K are given at two temperatures and apparent viscosities are shown at shear rates of 10 s^{-1}, 50 s^{-1} and 100 s^{-1} for LAS pastes.

Table 16 Alkylbenzene sulphonate (LAS) paste viscosity at various AD concentrations and temperatures

Type of AD	conc %	K Ns/m^2		viscosities Ns/m^2 at shear rate LAS					
		$1\ s^{-1}$		$10\ s^{-1}$		$50\ s^{-1}$		$100\ s^{-1}$	
		40°C	60°C	40°C	60°C	40°C	60°C	40°C	60°C
Dob 055	40	20	15	3.4	2.4	0.95	0.66	0.54	0.4
Korenyl	43	24	17	4.6	3.0	1.4	0.95	0.85	0.6
SiR-X-12L	51	77	48	13.0	8.0	3.8	2.2	2.2	1.2

The power law constant, n, for the 3 examples at 40°C and 60°C is about 0.2. Table 16 shows clearly how dramatically the apparent viscosity drops with increasing shear.

The following example shows the practical implication of this pseudoplastic behaviour. The shear rate in pipeline systems transferring paste can be estimated from the laminar flow pattern:

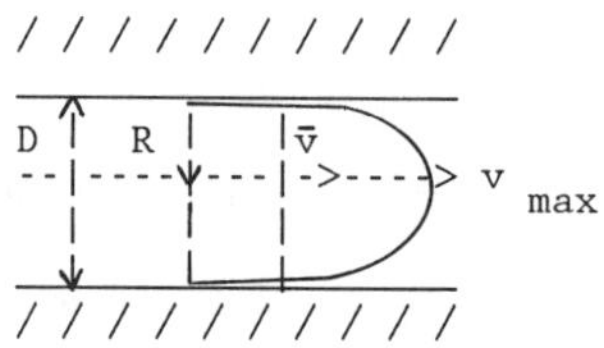

In laminar flow of pseudoplastics the flow velocity pattern is by approximation parabolic and v_{max} in the centre of a tube is about 2 x v. Assuming a paste velocity of say 0.5 m/s in a pipeline of 3" diameter the shear rate $\Delta v/\Delta y$ can be estimated

v_{centre} would be $2 \times \bar{v} = 1$ m/s

Δy equals R, thus $\Delta v/\Delta y = 1/0.0375 = 27\ sec^{-1}$

This would result in an apparent viscosity for Korenyl paste at 43% AD, 40°C:

$$\eta = Kj^{n-1} = 24 \times (27)^{-0.8} = 1.7\ Ns/m^2$$

The pressure drop over a pipe length L can be expressed, for Newtonian laminar flow by:

$$\Delta P = 32\ \eta Lv/D^2 \quad \text{(Poiseuille)}$$

When we substitute η, the viscosity for a Newtonian liquid, by the apparent viscosity for AD pastes, the calculated ΔP values will be roughly correct. For the above example ΔP over 20 m pipe length would become with the Poiseuille equation $1.0 \times 10^5\ N/m^2$ or 1 Bar. With the more precise relation for pseudoplastic flow the calculated pressure drop would have been 0.7 bar. The Poiseuille equation presents a conservative estimate for pseudoplastic liquids.

To quench the heat of the neutralisation reaction, all suppliers use loop type circulation reactors, consisting of a circulation pump with a capacity of about 10 times the volumetric flow of incoming liquor (organic acid and diluted alkaline liquor) and outgoing paste, a high shear homogeniser and a shell-and-tube heat exchanger to cool the circulating mass.

In the high-shear homogeniser the apparent viscosity of the paste will be minimal, resulting in an instantaneous mixing and reaction of organic acid and diluted caustic in the paste carrier. The circulating mass is used as a heat sink for the exothermic reaction. Due to this effect the temperature increase after the homogeniser will only be in the order of 5°C. Typical temperatures would be 50°C before the homogeniser, 55°C after reaction at the exit of the homogeniser, 50°C post heat exchanger, thus being the temperature of the paste leaving the loop reactor and the paste circulating in the loop reactor.
The loop circulating reactor outlined above expresses in chemical engineering terms, the characteristics of a perfect mixer. This type of reactor presents the following characteristics required for the neutralisation step:

(i) high efficiency in distributing the acid phase and diluted caustic in the main stream of paste and mixing the components on a micro-scale, so that the neutralisation reaction will be completed in a very short time;

(ii) good temperature control by using the heat-sink capacity of the circulating paste cooled by an heat exchanger which forms part of the loop reactor;

(iii) a small hold-up system with a quick system response, important for good pH control.

For the manufacturer of 70% AD SLES and 70-75% AD coco-PAS, rheological changes that take place during the neutralisation stage start-up can result in very high apparent viscosities (K is about 100 Ns/m^2 or 100 000 cP) and thus high pressure drops in the neutralisation loop reactor. This has implications for the mechanical design of pumps, mixers, pipe sizing and equipment lay-out.

The technique for high-active detergent paste manufacture exploits viscosity minima that occur at active levels of 70% AD for SLES and 75% AD for coco-PAS. However, during the start-up of the neutralisation loop, active levels of less than 70% and 75% occur. At these reduced active levels the viscosity of the active paste goes through maxima which can be reduced through the accurate control of excess caustic soda or sodium sulphate levels. Normally, controlling excess caustic levels (0.1 to 0.2%) is preferred since caustic soda is the commonly used neutralising medium.

Owing to the high viscosities and the need for careful control of the relative amounts of organic acid and caustic soda, achieving good mixing is difficult but of great importance. Localised pockets of un-neutralised acid can result in acid hydrolysis of the product.

The specified circulation pump operating pressure can be as high as 10 bar with a suction pressure of 2 bar and a maximum design pressure on the loop of 30 bar. Ballestra, for example, recommend a two-speed mixer and recycle pump to run at reduced capacity and circulation rate during the neutralisation loop start-up phase (when viscosities are at their highest value) to avoid excessive pressure in the loop.

From the steep titration curve of a strong acid with a strong diluted alkaline phase, it will be evident that the slightest adjustment of the amount of caustic will result in a major change in pH value. Acid paste must be avoided at all times, if only to prevent corrosion. Because of the steep pH curve, the recommendations for neutralisation control are as follows:

(i) dose the caustic soda (NaOH) diluted, i.e. 10 to 20% depending on required paste strength. In other words, the commercial 46%-50% caustic should be diluted to required concentration before dosing to the neutralisation loop. Small variations in caustic demand can be made more accurately with dilute than with concentrated solutions of caustic;

(ii) split the caustic dosing into two branches: use flow control for the main quantity (95%) of the caustic required for neutralisation, with the remaining part (5%) being controlled by reference to the desired pH value of the paste;

(iii) the actual pH measurement should preferably be made in a separate diluted paste sample to avoid electrodes fouling with thick paste and to prevent electrode deterioration;

(iv) a weekly inspection and calibration of the pH instrument is essential and should be carried out by a qualified instrument specialist.

5.6.2 The Ballestra stirred-tank neutraliser system

The neutralisation takes place in a stirred tank reactor with cooling coils to which the sulphonic acid (ex cascade sulphonation reactor train or from storage) and the diluted caustic are continuously added. The stirred-tank neutraliser is the final reactor in the total train of seven (4 sulphonation reactors, 1 ageing vessel, 1 hydrolyser reactor and last, but not least, the neutralisation reactor).

The sulphonic acid is introduced beneath the paste surface, near the propeller-type agitator. Through a pipe manifold process water and caustic are introduced beneath the paste surface, sufficiently far away from the impeller to avoid it being corroded due to strong pH fluctuations.

The main liquid streams are added via piston-type pumps: one pump-head for caustic (up to 95% of required amount) and one for process water. Two smaller capacity pumps are used for pH control, one feeding a small side-loop of paste with dilution water where the pH of the diluted stream (15-20% AD) is measured, while the other feeds a small caustic stream for pH adjustment.

The internal pH loop consists of a gear-pump, an in-line mixer for paste dilution and an in-line pH meter/control unit that is connected to the small capacity caustic dosing pump.

The paste volume (heat sink) in this vessel related to throughput corresponds to a mean residence time of about 15 minutes. The system can handle LABS pastes up to 55-60% AD. Bleaching facilities (branched AB from CSTR system) via either direct NaOCl addition or in-situ reaction between Cl_2 and dilute caustic, are available and can be delivered by the Ballestra Group. Normally there is a finishing tank where quality can be checked before transfer to paste storage tanks in the tank farm. The agitated tank reactor for neutralisation can be seen as an "internal loop reactor", the agitator acting as the homogeniser and circulation pump. The cooling coils are the heat exchanger and the paste volume acts as the heat sink.

5.6.3 Ballestra double-step neutralisation system (Neutrex) (see fig. 33)

For Ballestra plants running a wider range of products like alkylbenzenes, alcohols and ethoxylated alcohols, the double-step neutralisation system is recommended. This system allows neutralisation to occur in two steps: 95% of the neutralisation occurs in the first loop reactor, always working under strong caustic conditions; and neutralisation is completed in the compact intense mixer second step, in series with the first. This procedure may be followed for LAB-SA neutralisation.

For alcohols and alcohol ether feedstocks, complete neutralisation takes place in the first loop and buffering agents/preservative can be added in the compact intense mixer second step unit.

The first neutralisation step consists of a recirculation loop comprising an agitated in-line mixer, a shell-and-tube heat exchanger (both cooled with cooling water) and a gear-type or mono-type recirculation pump. 95% of the LAB organic acid is sent to the first loop reactor, together with the full amount of caustic/dilution water calculated for 100% LAB acid, i.e. neutralisation in the first loop reactor is carried out with an excess of 5% alkali. From the first loop reactor the alkaline paste flows to the second reactor, a small stirred-tank reactor with cooling mantle, where the rest of the sulphonic acid (5%) can be added to complete the neutralisation. Besides the continuous dosing of the acid there is the possibility of dosing extra alkali to the second step for pH correction, if necessary.

Final pH control takes place in the second compact reactor, promoting a quick system response. The first loop reactor also has pH measurement, which in certain circumstances can be used for control, for example when it is required to introduce all the acid in the first loop (alcohol ethoxylates).

The double-step neutralisation system is suitable for the production of 70% SLES. Any type of organic acid in combination with any type of neutralisation agent can be used (except stoichiometric amounts of Na_2CO_3).

In order to change over from 27% SLES to 70% SLES in the Ballestra Neutrex system (see also figure 47), the following procedure is recommended (see also figure 33):

1) close the cooling water valves on the heat exchanger to increase the temperature;

2) divert all the caustic to the in-line mixer first loop;

3) shut off the process water to promote paste concentration;

4) when the pressure begins to increase, reduce the speed on two-speed circulation pumps, thereby reducing the pressure;

5) as soon as the pressure peak has been reached and the pressure begins to reduce, switch circulation pump back to high speed;

6) adjust the cooling and process water to the desired level.

Polyglycol or ethyl glycol can be added to fatty alcohol sulphates at 70% concentration to lower the viscosity, if required.

Figure 33 Ballestra double step neutraliser

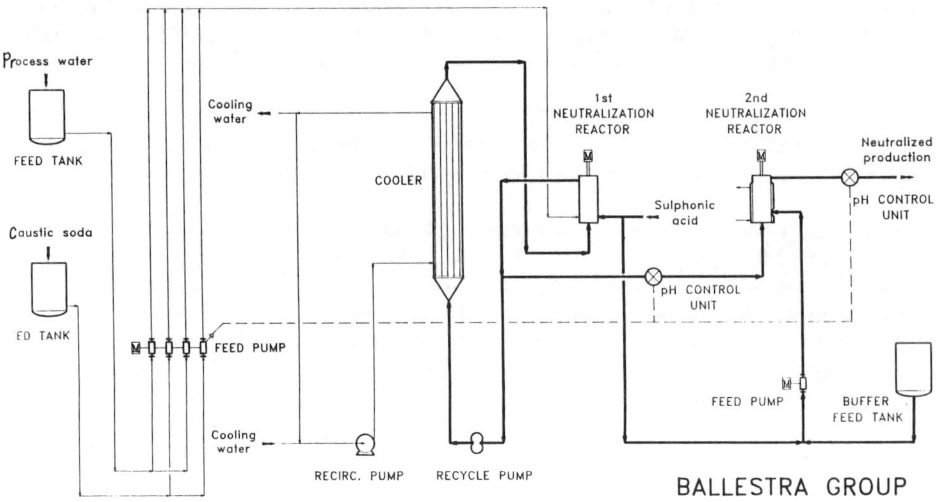

5.6.4 The neutralisation equipment of MM, Mazzoni and Chemithon (see fig 34, 35 and 36)

The three systems are similar and therefore are not discussed separately. The circulating loop reactor consists of a gear-type circulating pump with variable throughput, a shell-and-tube heat exchanger and a homogeniser. In the Chemithon plant the homogeniser is a kind of centrifugal pump, while MM and Mazzoni use a high-shear rotor/stator homogeniser. In the Chemithon plant the homogeniser is either a modified centrifugal pump or, for highly active products a high shear mechanical mixer.

The neutralising agent and process water are fed to the homogeniser via a flow control system. The alkali stream is normally split in two streams: a main stream (90-95% of required amount) and a secondary stream connected to the paste pH control (see 5.6.1.).

The paste pH is measured by an in-line pH meter in the loop just after the homogeniser (MM and Mazzoni). Chemithon takes a small paste side-stream from the loop, diluted with water via a static in-line mixer, and after pH measurement in-line, the diluted stream is re-introduced into the loop. This pH meter controls the flow of the neutralising agent and consequently the pH of the paste in the loop.
The loop reactors operate under pressure (1-3 bar depending on feedstock and paste

concentration, although during start-up higher pressure may occur) which is regulated by a pressure controller discharge valve. The recirculation-to-output ratio is about 10 to 15. The mean residence time in the loop reactor is about 5 minutes.

On the cooling water side, a loop system is normally used. Fresh cold cooling water (10-20°C) is supplied to a recirculating stream of cooling water (about 30°C) in which the amount supplied is regulated by a temperature indicator/ controller in the cooling water loop. To avoid skin formation of paste on the inside tube surfaces, which would seriously hamper heat transfer, the differential between the cooling water temperature and the neutralisation temperature is kept close to 20°C.

The homogenisers can suffer corrosion due to pH fluctuations and high paste concentrations. Extra attention when ordering equipment is required to ensure that the correct materials of construction are used.

All equipment allows production of 40% active alkylbenzene sulphonate and 30% active sodium lauryl alcohol sulphate products. Normally, the equipment also allows production of other products such as high active (50-60%) alkylbenzene sulphonate and 70% AD ethoxylated alcohol sulphates. Specially designed neutralisers of this kind yield products such as 75% AD sodium coconut alcohol sulphate. All these options also hold for the Ballestra Neutrex system discussed in section 5.6.3.

Standard loop neutralisers use caustic soda solution as neutralising agent. Other agents such as aqueous or anhydrous ammonia, caustic potash, or ethanolamine can also be used. In addition liquid additives such as ethyl alcohol and urea solutions will blend with the paste in the loop reactors.

Figure 34 Mazzoni: Sulphonic acid neutralisation

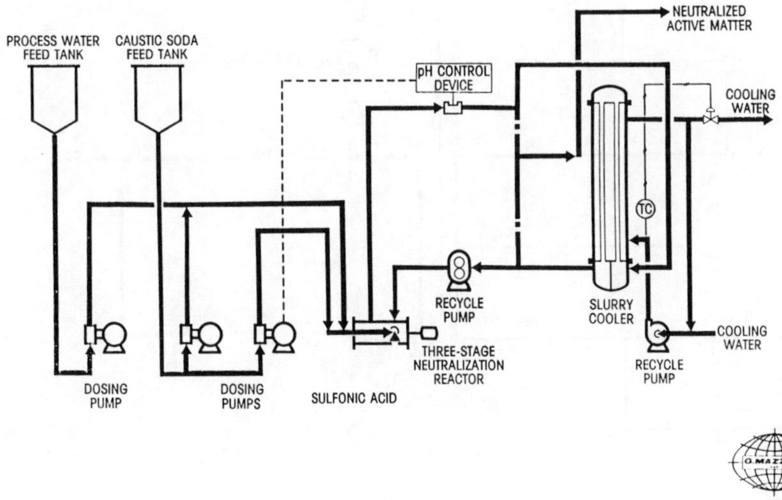

Figure 35 Mazzoni: Three stage neutralisation reactor

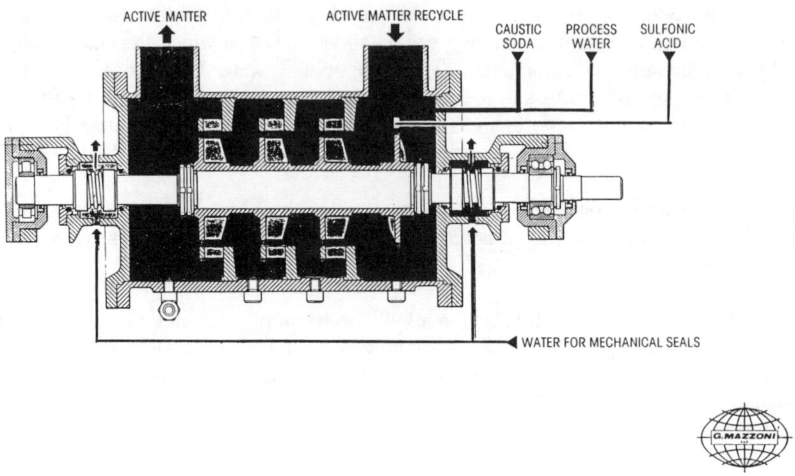

Figure 36 Chemithon: Neutralisation system

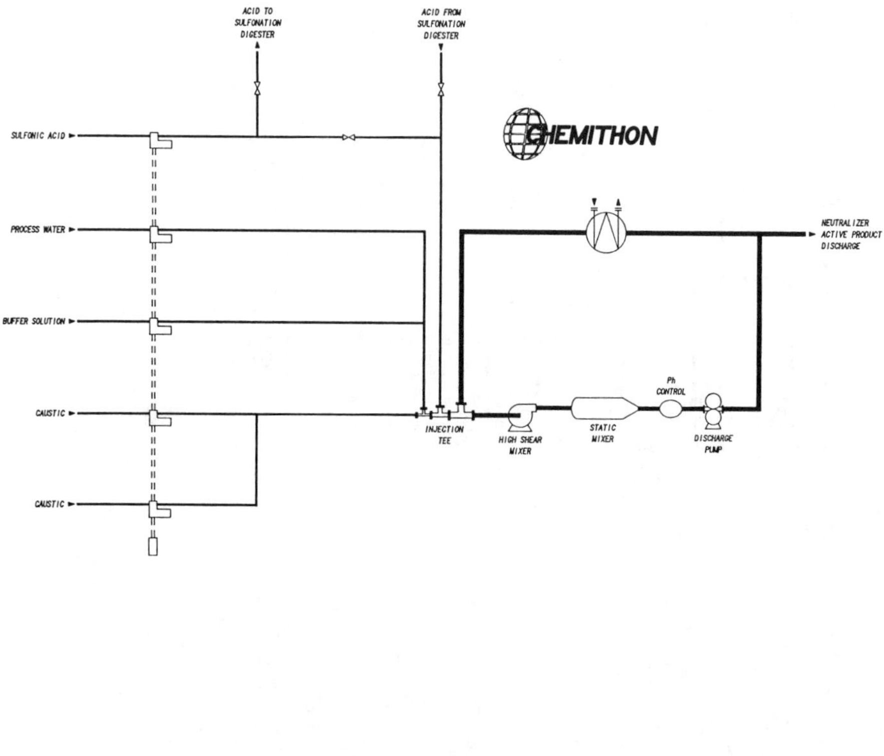

5.7 Chlorine dosing equipment for in-situ manufacture of diluted NaOCl solutions

The Wallace and Tiernan chlorinator type V-741-G, with a capacity of 9.5 kg Cl_2/h dosing feed range chlorine gas 1:20 and accuracy 4% of indicated feed rate is used by Ballestra (see fig. 37). It is an example of a well-designed plant for introducing chlorinated water to the 50% caustic/dilution water manifold. The operating vacuum is developed by an aspirator-type injector. As soon as the operating vacuum is developed, the chlorine gas enters the chlorinator through the combination valve. This valve closes if the vacuum decreases to an insufficient level or vents to the atmosphere if over-pressure occurs. It also maintains the suitable vacuum-level ahead of the V-notch variable nozzle.

The chlorine gas flow-rate can be read conveniently on the large scale flowmeter. The scale permits the flow-rate to be read with high accuracy. This chlorine flow-rate is set by manually adjusting the V-notch variable orifice.

The vacuum regulation valve produces a constant vacuum after the V-notch orifice. An incorporated vacuum breaker prevents excessive vacuum. At the injector the chlorine gas is dissolved in the caustic dilution water and thus mixed thoroughly. The resultant solution is discharged in the manifold where dilution water and caustic solution are mixed.

The unit has been designed for wall mounting. An extremely careful selection of material provides for long-term corrosion resistance and trouble-free operation.

The chlorine can be dissolved in the caustic dilution water. The advantages of this system are:

(i) there will be no chlorine gas pipeline in the sulphonation plant itself. Chlorine dissolved in water will be transferred;
(ii) there is no heat effect during the absorption of chlorine in water and therefore no heat exchanger is required;
(iii) the system is simple, safe and compact in design. It is widely used for water treatment.

In table 17 the amount of caustic dilution water and chlorine are related to Ballestra Sulphurex CSTR plant capacities.

Table 17 In-situ hypochlorite production data

Plant cap. in kg 100% AD/h	Caustic dilution water assuming 40% caustic from stock and 45% AD paste kg/h	Required kg Cl_2/h	Conc. in water kg/m^3	NaOCl conc. wt% in diluted caustic
1000	1015	2	2	0.15
2000	2030	4	2	0.15
3000	3045	6	2	0.15
5000	5075	10	2	0.15

The 0.15% strength NaOCl solution ultimately formed in situ is about 1% of the strength of commercially obtained NaOCl solutions (6-15%).

Figure 37 Chlorine dosing equipment

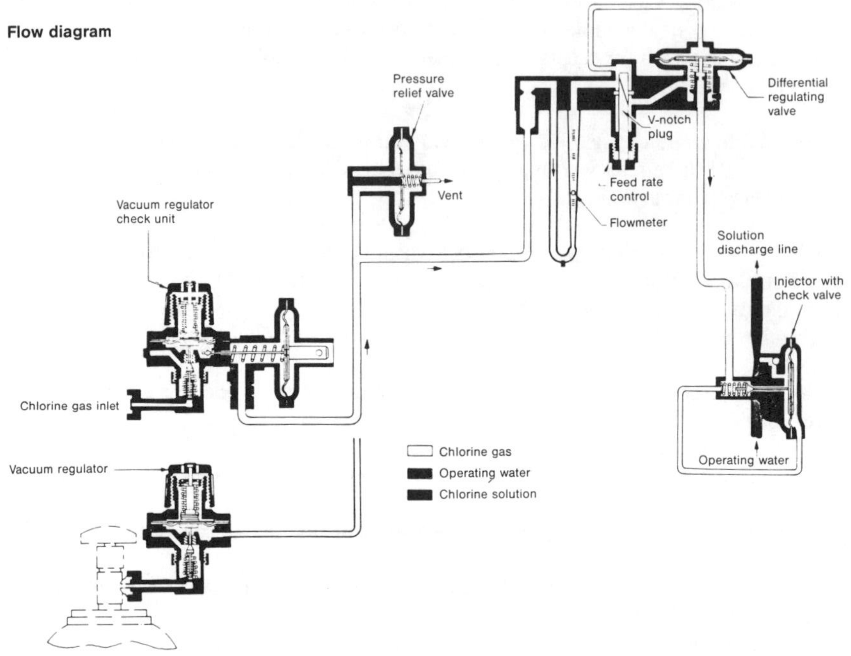

5.8 Sultone hydrolysis equipment (see fig. 38)

The process flow diagram is shown in figure 38. Alpha-olefin sulphonate from the neutralisation section is transferred by a booster pump, providing the required pressure to avoid water evaporation in the system, to the pre-heater and hydrolyser itself. The fresh olefin sulphonate flows through the pre-heater counter-current to the hydrolysed material coming from the hydrolyser. This system results in a heat recovery of 75-80%. Two additional small heat exchangers are required: a heater between the pre-heater and the hydrolyser to fine-tune the temperature before the alpha-olefin sulphonate enters the hydrolyser and to pre-heat the flow during start-up. Finally, a cooler to cool the product flow leaving the hydrolyser is installed.

Sultones, especially the D-1,4-alkane sultones, hydrolyse slowly below 100°C. Hydrolysis takes 20 minutes at 150-160°C (see 4.3.4.). The hydrolysis reactor approaches a plug flow characteristic. The whole hydrolysis operation is carried out at a pressure slightly above the saturated water vapour pressure that corresponds to the chosen temperature e.g. 170-180°C -->

9-10 bar. Essentially similar continuous hydrolysis systems have been developed by all sulphonation plant suppliers.

5.9 Exhaust gas cleaning equipment

Unconverted SO_2, SO_3 (mainly as sulphuric acid mist) and entrained organic acid are present in the exhaust gas and are the main potential atmospheric pollutants. The recommended practice is to use an electrostatic precipitator (ESP) to trap any particulate material in the form of mist or aerosols and a caustic soda scrubber which absorbs SO_2 and small amounts gaseous SO_3 that will pass the ESP.

Figure 38 Alpha olefin hydrolysis equipment

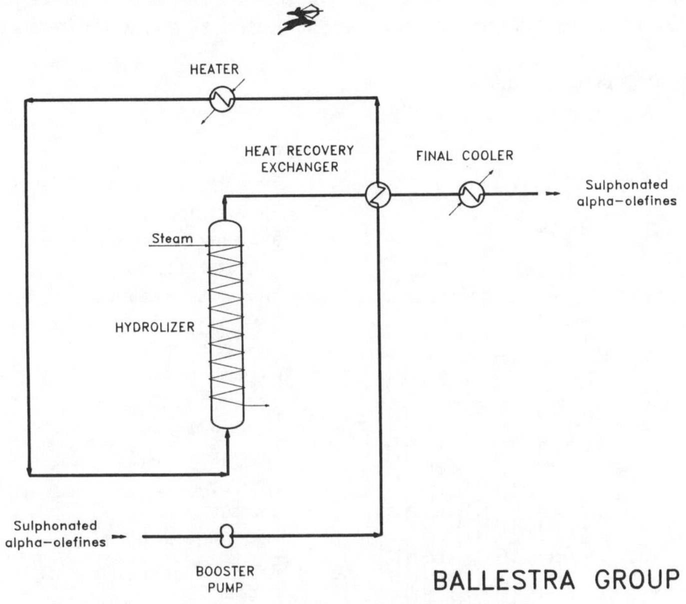

5.9.1 The electrostatic precipitator (ESP) (see fig. 39)

The function of the ESP is to clean a gas stream via the removal of particulate matter under the influence of electrostatic forces generated between two electrodes, a wire and a tube.

The contaminated gas stream is evenly distributed to a number of tubes through which it passes, coming under the influence of the electrostatic field generated between the tubes and centrally hanging wires (figure 40).

The principles of electrostatic precipitation are well-known. An electric field within a small band around a discharge electrode (the wire) is of sufficient strength to accelerate free electrons in a gas to a high-energy state. The high-energy electrons ionize the gas and produce electronegative ions which migrate away from the negative electrode. These ions intercept and become attached to the particulates entrained in the gas stream, as a result of which the particulates attain a high negative charge and are subjected to strong electrostatic forces. These forces cause the particulates to migrate toward the positive anode (the tube), where they coalesce on the surface and drain off under the influence of gravity.

Within the ESP the following actions occur:

(i) the organic/sulphuric acid droplets are ionized and collected on the tube walls. This is the primary function of the ESP;

(ii) a limited amount of SO_2 gas and SO_3 gas is removed by physical absorption into the organic liquid film on the walls of the tubes.

It is clear that the ESP cannot remove dry SO_3 gas and the moisture available to convert SO_3 gas to SO_3/H_2SO_4 mist originates from residual water in the process air and water from sulphur.

Figure 39 ESP Construction details

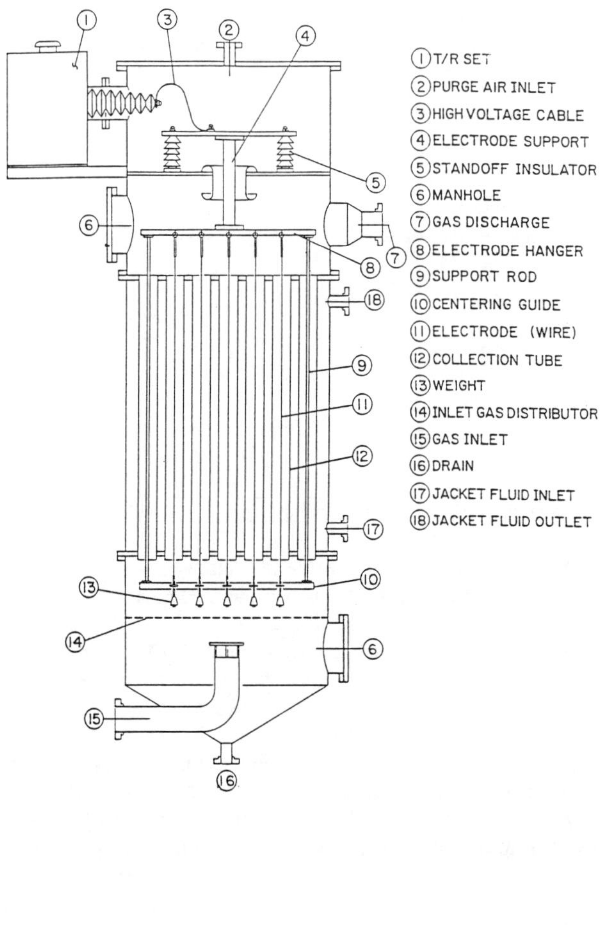

Figure 40 Details of wire hanging mechanisms

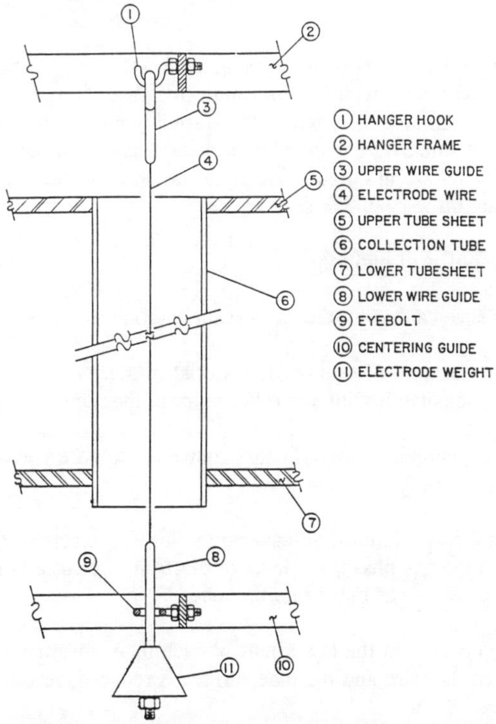

The quantity of residue obtained from an ESP is primarily dependent upon the organic feedstock being used and its quality. Suppliers suggest that the following levels of residues would be achieved in practice although the guaranteed levels from Mazzoni and MM are only 0.5%.

alkylbenzene feedstock	0.1-0.2% of organic feedstock
alcohol ethoxylate feedstock	0.2-0.3% of organic feedstock

Each tube and wire combination within an ESP should be considered as an ESP in its own right. An ESP is therefore a collection of identical and individual ESPs working in parallel. The key parameter involved in the design of an ESP is the field strength generated on the electrode and collecting surface of an individual tube and wire combination. An increase in field strength leads to the possibilities of:

(i) increased cleaning efficiency if velocity and tube dimensions remain constant;

(ii) increasing the gas velocity through the tubes, maintaining the same operating performance;

(iii) decreasing the tube length required to maintain the same operating performance.

In practice there is a wide range of electrical operating conditions. Suppliers recommend 16-25 kV, 30-70 mA.

The gas velocity in individual tubes is in the range of 0.3-0.4 m/s. Tube diameters are of the order of 160-170 mm, with an effective tube length of 2-3 m. Wire material and wire diameter vary widely. Mild steel, stainless steel wire 316, Hastelloy and even Incoloy are used, diameters varying between 2 and 4 mm. Wire construction material is varied due to wire breakage experiences. Wire breakage occurs predominantly at the base of the electrode wire, an area which has the following three characteristics:

(i) the highest level of wall build-up;

(ii) high gas turbulence caused by the gas stream entering the tube;

(iii) the highest concentration of removable material in the gas stream, in comparison with any other point in the tube.

The following theory explains how these factors contribute to wire corrosion caused by electrostatic forces:

- since there is a direct relationship between the level of electron discharge and particulate concentration in the gas-phase, the level of electron discharge (wire to tube wall) is greatest at the point of gas entry into the tube;

- as wall build-up occurs in the lower part of each tube, forming a conductive layer, the distance between the wire and the tube surface is reduced, leading to a greater electron discharge;

- as wall build-up increases and the distance between wire and wall decreases, local arcing may happen resulting in a continual migration of metal ions from the wire, the net effect being a reduction in the wire thickness;

- wires under even a small tension vibrate when subjected to a high potential DC (the higher the voltage, the higher the amplitude of vibration) which, as the wire thickness reduces, leads to the possibility of vibrational fatigue and eventually wire breakage.

In order to reduce the frequency with which wire breakage may occur, the following steps can be taken:

(i) make sure that wires are absolutely centrally positioned in tubes and maintain even gas distribution across tube nest;

(ii) use intermediate cleaning to reduce wall build-up;

(iii) eliminate the exposed length of wire outside the tube by increasing the teflon wire coverage, so it enters into the lower part of each tube (see fig. 41);

(iv) inspect wires annually and replace all (mild steel) wires once a year. Mild steel wires are cheap and easy to handle, weld and correct.

Since an ESP is in fact a collection of individual ESPs it is critical that the total duty placed on the unit is distributed evenly across the individual tube and wire combinations. This point is strongly emphasised by suppliers, all of whom have taken steps to improve gas distribution in their installations.

Figure 41 Points of wire failing

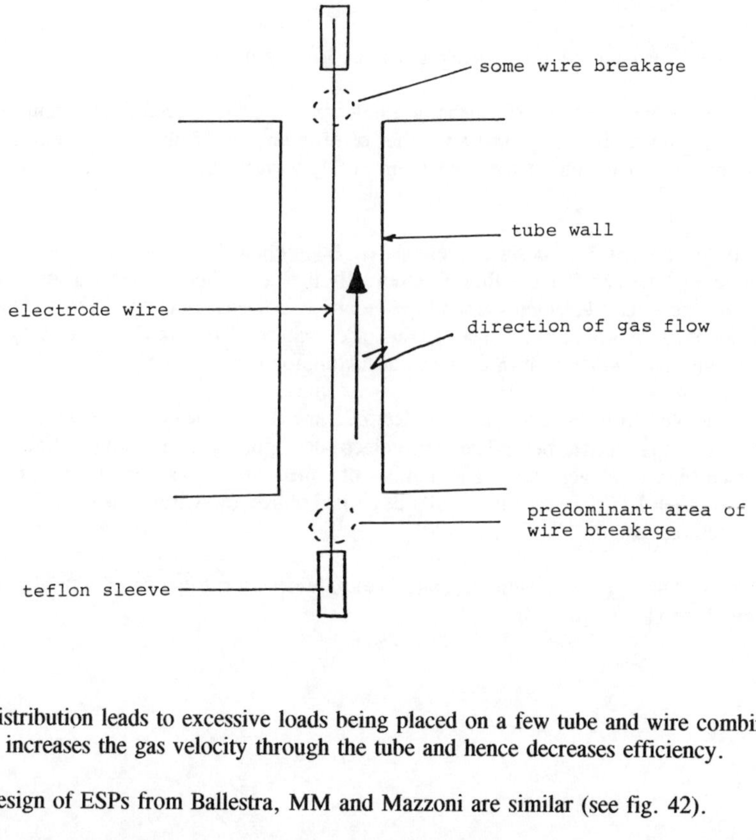

Bad distribution leads to excessive loads being placed on a few tube and wire combinations, which increases the gas velocity through the tube and hence decreases efficiency.

The design of ESPs from Ballestra, MM and Mazzoni are similar (see fig. 42).

In the Ballestra design the tubes are connected at their upper end to a plate which seals the ESP into two halves, between which gas can only pass via the tubes. The gas enters the ESP halfway up the shell housing the tube nest. This arrangement of exhaust gas entry performs three functions:

(i) moderately heating the tubes (the exhaust gas post sulphonation plant is about 40-50 °C) by circulation around them. This helps to keep the organic material mobile on the inner wall of the tubes;

(ii) a cyclone separation effect, removing some organic/liquid material from the gas phase in the shell prior to entering the tubes;

(iii) generating an even distribution of the gas to the individual tubes, since gas flow is induced by the pressure drop across the tubes, and not by the directional momentum of the gas stream.

The MM design is along the same lines as Ballestra, the only difference being that the gas enters the ESP higher up.

The Mazzoni design has the following features:

(i) heating mantle around the tubes, using air as the heating medium;

(ii) gas entry at the base of the ESP, using a double grid to achieve even distribution to all the tubes. Each grid is made up of a number of different-sized holes, which deflect the natural vertical momentum of the gas stream evenly towards the central and peripheral tubes.

Figure 43 illustrates the new ESP design developed by Chemithon. Much of the vessel design remains the same e.g. the path for gas flow is essentially the same. The principal distinctions are the electrodes and the electrode support assembly. The high-intensity ionizing electrode is a disk shape, the outside edge of which has a small radius of curvature. The disk electrodes are supported by a rigid mast which in turn connects to a common supporting plate.

The corona is generated from the edge of each electrode and the geometry enables precise location of the collecting electric field. The entire electrode support is rigid, which allows operation at much higher gas velocities. The number of collection tubes required for a given quantity of gas is much less than for the 3 ESPs described above and wire breakage is eliminated in the new Chemithon design.

In table 18 some features of the Chemithon Disk Electrode design are compared with the traditional wire electrode design.

Figure 42 A few examples of ESP's from different sulphonation equipment suppliers

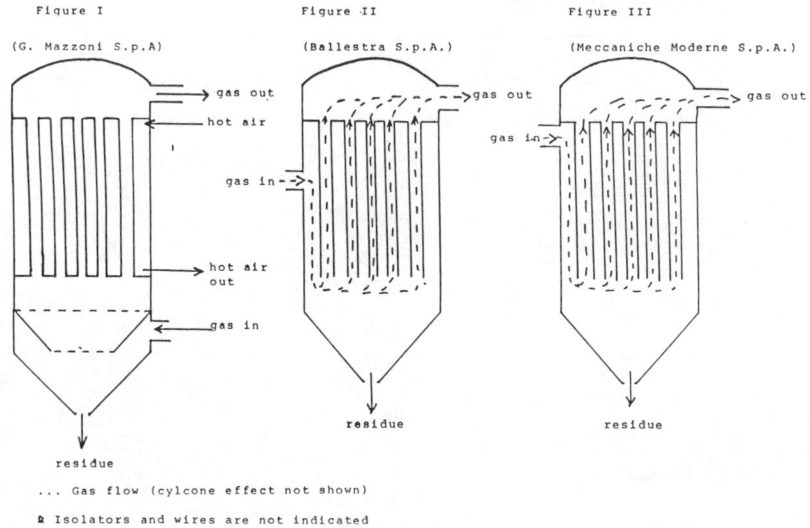

Figure 43 Chemithon's Disk Electrode ESP design

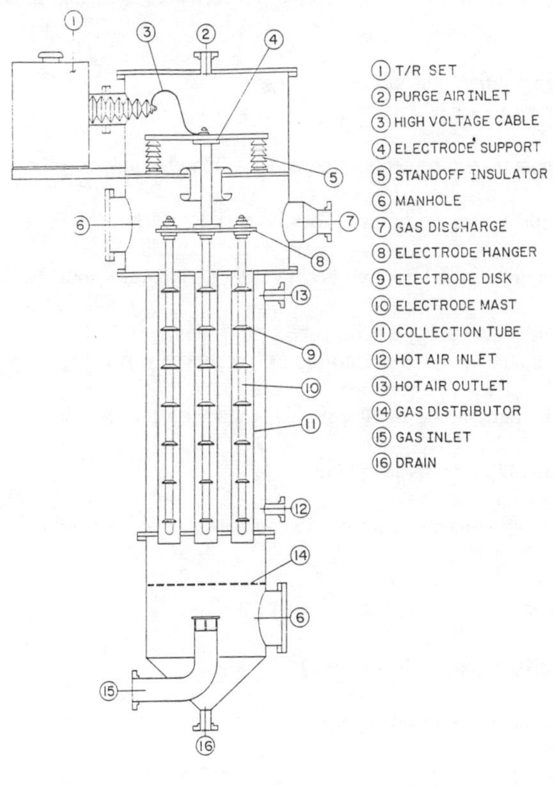

Table 18 Comparison of ESP performance at test plant during sulphonation of linear Tridecylbenzene (MW 262)

	wire electrode design	disk electrode design
Design factors		
inlet gas flow m^3/min	70.4	94.2
number of tubes	101	31
gas velocity per tube m/s	0.57	1.5
gap cm	7.9	6.0
Operating data		
AC Volts	325	440
AC Amps	22	18
DC kVolts	19	52
DC m-Amps	180	100
DC m-Amps per tube	1.78	3.23
Average field strength kV/cm	2.4	8.0

Chemithon claims efficiencies of 98-99% depending on feedstock type and quality. Efficiency is defined as follows:

$$\frac{\text{(particulate-in kg/h)} - \text{(particulate-out kg/h)}}{\text{particulate-in kg/h}} \times 100\%$$

There are a few fundamental requirements to consider when cleaning an ESP:

(i) the unit should not be opened unless it has been properly grounded;

(ii) any part of the equipment which will be in contact with water must be earthed;

(iii) cleaning with cold water using a hose is recommended.
Washing water should be directed into the SO_2 scrubber water pit;

(iv) after washing, the unit should be thoroughly dried with process air.

The total operation normally takes about 4 hours.

In conclusion, the following conditions are of vital importance for successful operation of ESPs in sulphonation plants:

- ensure good gas distribution across parallel ESP tubes;
- maintain high voltage across electrodes (16-25 kV);
- avoid arcing (premature wire breakage);

- dry thoroughly after washing;

- ensure that purge air to the electrode set is at the correct flow-rate and temperature to avoid fouling.

5.9.2 SO_2 removal from exhaust gas, the caustic scrubber (see fig. 44) and sulphite oxidation

Ideally the feed to the scrubber should consist of

- traces of dry SO_3 gas
- SO_2 gas not converted to SO_3
- "air".

Within the scrubber the following occurs:

The SO_2 gas is removed to the required level by reaction with dilute caustic soda solution. The chemical reaction resulting in sodium sulphites, is instantaneous under alkaline conditions (pH 8-9) and the rate of reaction is gas-phase controlled (see 4.5). Small amounts of dry SO_3 gas will be removed by reacting with the dilute caustic soda solution, forming sodium sulphate.

With small amounts of SO_3, its vapour pressure in the feed is very low and chemical absorption is the favoured reaction, competing with H_2SO_4 mist formation. The caustic soda scrubber will not remove dry SO_3 in large concentration as H_2SO_4 mist prevails. This H_2SO_4 mist passes through the scrubber without reacting with the caustic. The aerosol is so finely divided in the gas phase that it apparently can pass the scrubber without reaching the caustic soda solution interface.

The absorption efficiency of the scrubber is dramatically reduced by the presence of any active detergent. The theory is that the active forms a boundary layer between the gas-phase and the caustic, preventing or at least limiting contact. Also, foam formation will hamper normal operation of the scrubber. Hence for the efficient operation of the scrubber system, the ESP's primary function is to prevent the active contamination of the scrubber liquor, i.e. all organic acid mist and droplets should be trapped in the ESP.

The exhaust gas leaving the ESP, virtually free from any organic acid and sulphuric acid mist, enters the SO_2 scrubber made of acid-proof construction material. The caustic scrubber equipment consists of:

- a packed bed section filled with a packing material e.g. Raschig rings of 1";

- a demister in the top section to trap small gas-entrained liquid droplets;

- a caustic solution recirculation pump, with flowmeter to check recirculation flow rate;

- a control system for the addition of the dilute caustic soda required to convert the SO_2 to Na_2SO_3;

- a pH control device in the recycle solution of the scrubber system, controlling the amount of caustic to be fed to the scrubbing system.

The pH and the inorganic salt concentration in the circulating liquid are kept to the pre-set value by means of pH control and inorganic strength (10% salts w/w) control (dilution water feed system). These controls are very important in order to keep the alkalinity at the required pH value (10-12) and to avoid crystallization of the inorganic salt from solution.

There are differences between suppliers. In the Mazzoni design, the gas stream leaving the ESP is mixed in a venturi with the circulating sodium sulphite solution, which is sprayed via a nozzle co-currently into the gas stream, followed by a subsequent two-stage packed bed counter-current absorption unit. The packed bed is split in two compartments to make sure that proper distribution over the packing material takes place. MM's scrubbing system is very similar to Mazzoni's design. It consists of a vessel having a column with two beds filled with packing material fitted on it. The construction material is fibreglass-reinforced polyester. The Ballestra system has only one packed bed, normally fitted with 1.0" PALL Rings. All suppliers have a demister section in the top of the column to trap entrained liquor, which flows down to the absorption part of the column.

The absorption of SO_2 (and traces SO_3) is gas-phase controlled. The HTU (Height of a Transfer Unit) for gas-phase controlled absorption is about 0.3m, assuming 1.0" Raschig Rings and proper wetting of the packing material (correct liquid circulation rate and proper liquid distribution over the bed cross-sectional area). The required bed height can be calculated from the simple absorption equation.

$$\ln Co/Ce = H/HTU$$

in which

Co	=	SO_2 concentration in gas entering scrubber system (mg/m^3)
Ce	=	SO_2 concentration in gas leaving scrubber system (mg/m^3). This value should be 5 mg/m^3 max.
H	=	is the total height (M) of the packed bed in the absorption column
HTU	=	Height of one transfer unit (M) In this example 0.3m for 1.0" Raschig Ring

The exhaust gas leaving the scrubbing system should normally have a residual SO_2 concentration of max. 5 mg/m^3; during plant start-up, when the conversion of $SO_2 \rightarrow SO_3$ will rise from 50% to 98%, the SO_2 concentration in the exhaust gas may be for a short period somewhat higher (10-15 mg SO_2/m^3).

Assume for example a plant of 1 ton AD/h requiring roughly 100 kg S/h. At a concentration of 4% SO_3 in air, the total amount of air would be at atmospheric pressure and 30°C 1980 m^3/h. At 98% $SO_2 \rightarrow SO_3$ conversion, the air entering the scrubber would contain 2060 mg SO_2/m^3 and at 50% conversion of SO_2 during start-up the SO_2 concentration would be 51500 mg/m^3 in the gas entering the scrubber. The required packing height can now be calculated for those two cases, assuming an exit SO_2 concentration in the exhaust gas of 5 mg/m^3:

98% conversion, $\ln 2060/5 = H/0.3 \rightarrow H = 1.8m$

50% conversion, ln 51500/5 = H/0.3 → H = 2.8m

For 1.0" Raschig Rings the packed bed should have a height of 2.5m, assuming a HTU value of 0.3m.

For a one ton AD/h plant (appr. 100 kg S/h), the following mass balance can be calculated for the caustic scrubber; assuming 98% SO_2 → SO_3 conversion during stationary running.

The amount of absorbed SO_2 is roughly 4.0 kg/h, requiring 5 kg 100% NaOH/h or 10 kg 50% NaOH/h; 7,875 kg Na_2SO_3 will be formed and 1.125 kg reaction water/h. Reaction equation:

$$SO_2 + 2\ NaOH \rightarrow Na_2SO_3 + H_2O$$

For a 10% sodium sulphite solution, the total amount of water would be 7.875 x 9 = 70.875 kg H_2O/h.

The required dilution water will become 70.875 minus 5.0 (water from 50% caustic) minus 1.125 (water from chemical reaction), totalling 64.750 kg water/h. A small amount of extra caustic has to be added to maintain a pH of about 12 in the liquor leaving the scrubbing system. pH 12 would correspond to a caustic concentration of 10^{-2} mole/l or kmole/m^3, corresponding to 40 x 10^{-2} = 0.4 kg NaOH/m^3. For a volumetric flow of about 0.071 m^3/h, the extra 50% caustic to be dosed would become 0.4 x 2 x 0.071 = 0.06 kg 50% NaOH/h.

The mass balance for the caustic scrubber is summarised in table 19.

Table 19 Mass balance for caustic scrubber for a 1 ton AD/h plant, corresponding to roughly 100 kg S/h, assuming 98% SO_2 → SO_3 conversion

Components	In (kg/h)	Out (kg/h)
SO_2	4.0	negligible
NaOH (50%)	10.06	-
NaOH (100%)	-	0.03
Dilution water in	64.75	-
Total water out	-	70.905
Na_2SO_3	-	7.875
Total kg/h	78.81	78.81

During plant start-up, the SO_2 → SO_3 conversion will be initially 50%, resulting in required quantities a factor 25 greater than indicated in table 18. However, the conversion will rise steeply with time. Within 10 minutes after start-up the SO_2 → SO_3 conversion will be 80-90%, assuming a proper preheating of the SO_2 - SO_3 converter reactor. The dosing system of caustic and dilution water to the scrubber can cope with SO_2 - SO_3 conversions from 80% upwards. During plant start-up procedures the caustic scrubber recycle-pump and caustic/water dosing will be started to accumulate sufficient excess of caustic to absorb the initial high SO_2 concentration after sulphur ignition.

In conclusion, proper SO_2 scrubbing, resulting in less than 5 mg SO_2/m^3 under all circumstances, requires the following conditions:

- avoid active detergent contamination of scrubber liquor;
- ensure good distribution of circulating alkaline liquor over cross-section of bed; circulation rate should be measured with in-line flowmeter;
- avoid pH drift and sulphite precipitation. The pH should be in the range of 10-12 and the dissolved salt concentration about 10 w/w%;
- check packing material height in column with supplier to ensure that low SO_2 emissions (below 5 mg/m^3) are guarenteed during normal plant operation and that the system can cope with much higher SO_2 concentrations during the start-up period.

Figure 44 Exhaust gas cleaning plant

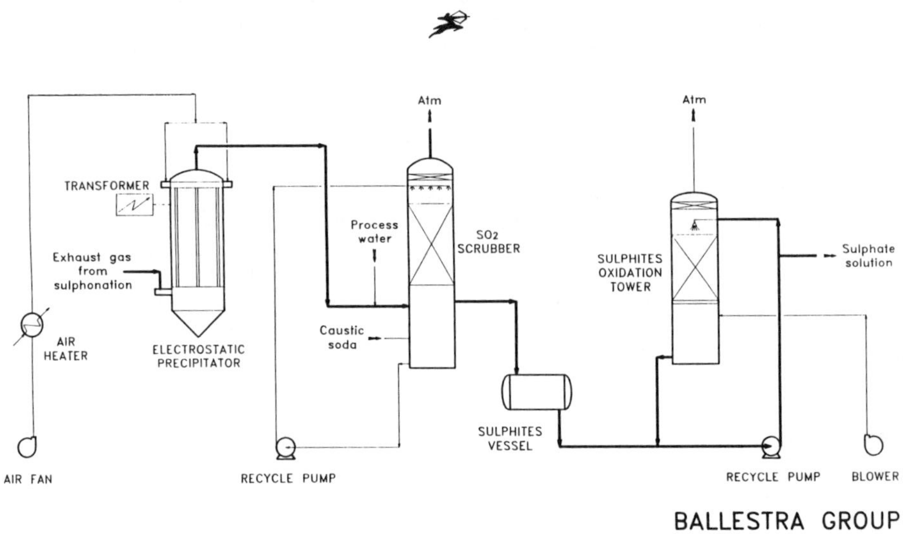

In Fig. 44 a sulphite/sulphate oxidation tower is depicted. Normally the sulphite/sulphate solution can be incorporated in the NSD detergent slurry before spray-drying takes place. When this is not possible, oxidation of the sodium sulphite solution leaving the scrubber may be required. The oxidation of sulphite to sulphate is carried out in the oxidation reactor, made of plastic material to avoid any corrosion. Atmospheric air is blown into the bottom of this unit and passes through a fixed bed made of special high surface area packing material. The sulphite-containing liquor from the SO_2 scrubber is circulated through this bed.

It may be necessary to add an oxidizing agent (hypochlorite) to oxidise completely any residual sulphite. The need for addition is normally established by off-line analysis.

5.9.3 Exhaust gas stack and composition of exhaust gas leaving the stack

Local authority regulations concerning the location, height and material of construction of the stack should be followed. Models to calculate the dispersion of gases from exhaust stacks are available if it is necessary to predict the SO_x concentration at points within and outside the factory boundaries. In most cases a PVC stack extending 5 m above the height of surrounding buildings is adequate.

The following are the recommended targets for the exhaust gas leaving the stack:

SO_2 max. 5 mg/m^3

SO_3/H_2SO_4 mist max. 10 mg/m^3

organic acid max. 20 mg/m^3

Sulphonation equipment suppliers accept these targets but would normally represent them as SO_x expressed as SO_3 below 20 mg/m^3.

In general a very light plume or no plume at all is observed from the stack of an SO_3/air sulphonation plant.

5.10 Bibliography

1. Ballestra SpA, (Via G. Fantoli 21/17, 20138 Milano, Italy), "Ballestra brochure" (1990).

2. Chemithon, (5430 West Marginal Way, S.W., Seattle, WA 98106, USA), "Sulfonation process equipment brochure".

3. G. Mazzoni SpA, (P.O. Box 421, 21052 Busto Arsizio, Italy), "SO_3 sulfonation/sulfation processing technology brochure" (1986).

4. Meccaniche Moderne, (Corso Sempione 51, 21052 - Busto Arsizio, Italy), "Sulfonation & Sulfation plants brochure", Catalogo All/8/278.

5. Lion Co., "T.O.-sulfonation/sulfation process". Info. Booklet 1978.

6. Herman de Groot W. (1989). "SO_3/air sulphonation reactor systems in practise". Paper presented at Practical Sulphonation Seminar, 18-20 April 1989, The Hague, The Netherlands. Center for Professional Advancement, P.O. Box H, East Brunswick, New Jersey 08816, USA.

7. Foster N.C. (1989). "The Chemithon sulfur burning gaseous SO_3 sulfonation technology".

Paper presented at Practical Suphonation Seminar, see reference 6.

8. Moretti G.F. (1984). "Sulphurex plant for continuous SO_3 film sulphonation and sulphation", Ballestra SpA, Milano, Italy.

9. Locatelli G. (Ballestra SpA, Milano) private communications.

10. Olivieri F., Corso D. (Mazzoni SpA, Busto Arsizio, Italy), private communications.

11. Johnson G.R., Crynes B.L. (1974). "Modeling of a thin-film sulfur trioxide sulfonation reactor"., Ind. Eng. Chem., Process Des. Develop. **13**, 6 (1974).

12. Davis E.J. et al. (1979). "An analysis of the falling-film gas-liquid reactor". Chem. Eng. Sci. 1979, vol. 34, pp 539-550.

13. Fok J.J., v.d. Velde S., Kooijman C. (1986). "Design and modeling of a falling film reactor for the sulfation of long chain alcohols", Groningen University, Dept. of Chem. Engineerimg, The Netherlands. (Private communications, graduation research project).

14. Nanninga R.M. (1989). "Modeling and simulation of falling film sulfonation". Groningen University, Dept. of Chem. Engineering, The Netherlands. (Private communications, graduation research project).

15. Gutierrez-Gonzalez J. et al. (1988). "Improved mathematical model for a falling-film sulfonation reactor". Ind. Eng. Chem. Res., vol. 27 (9) 1988, pp 1701-1707.

16. Coulson J.M., Richardson J.F. (1979). Chemical Engineering, vol. 3, (second edition), Pergamon Press, Oxford, UK. Chapter 6 "Non-Newtonian Technology", pp 425-450.

17. McKelvey J.M. (1962). Polymer Processing, John Wiley & Sons Inc. (Topic: Non-Newtonian flow).

18. Ballestra SpA, Milano. "Neutrex, Ballestra‘s double step neutralization plant". Brochure 1979.

19. Wallace & Tiernan (Postfach 149, D-8870 Gunzburg/Donau, W.Germany). Brochure: "Chlorator V-741-AG".

20. Levenspiel O. (1962). Chemical Reaction Engineering, Wiley International Edition, chapter 7, "Design for multiple reactors".

21. Falbe J. (1986). Surfactants in consumer products. Theory, technology and applications. Springer Verlag, Berlin, section 3.2.1.1. "General production technologies", pp 54-62.

22. UK patent GB 204306 7 B. "Ballestra film sulphonation in a multitubular reactor". Patent published 1 May 1983.

6 PRODUCT QUALITY, STORAGE, HANDLING AND SAFETY

6.1 Equipment suppliers guaranteed final product specification

The following group of tables states the final product specification guaranteed by the equipment suppliers, provided that raw materials of a specified, reputable quality are used and correct processing procedures and practices are strictly followed.
All tables assume neutralisation by caustic soda.

Table 20 Final product specification for Ballestra Sulphurex CSTR sulphonation system

FEEDSTOCK	free oil on 100% AD	sulphate on 100% AD	Klett* colour	AD conc. in paste %
A. Branched DDB Oronite 60 (tri-decylbenzene C_{13})	1.5	2.0	acid: 150-200 paste: 150	50 max.
B. Linear alkyl-benzene (LAB)	1.3	2.0	acid: 60 paste: 60	60-80 max.

* 5% AD solution, 40 mm cell N42 filter, unbleached.

Table 21 Final product specification for Ballestra multi-tube film reactor (MT-FFR)

FEEDSTOCK*	free oil on 110% AD 100% AD	sulphate on	Klett** colour	AD conc. in paste %
A. Branched DDB i.e. Oronite 60 (tri-decylbenzene C_{13})	1.5-2	2	acid: 100-150 paste: 100	50
B. Linear DDB	0.9-1.2	1.0-2.0	acid: 30 paste:30	60-70
C. Natural lauryl alcohol C_{12}-C_{14} C_{12}-C_{18}	1.0-2.0	1.0-2.0	5-10	30-75
D. Synthetic alcohol C_{12}-C_{14} range, for example: Alfol 12-14, Ethyl 12-14 or Dobanol 25	1.0-1.5	1.0-2.0	5-10	30-75
E. Alcohol nat. (type Dehidol LS-2) or synth. (type Dobanol 25/3), ethox. with 2-3 EO	1.0-1.5	1.0-1.5	10-20	27-70
F. Natural fatty alcohol C_{16}-C_{18}	2.0-3.0	1.5-2.5	30-50	30-60
G. Synthetic alcohol C_{16}-C_{18}, for example: Alfol C_{16}-C_{18}	2.0-3.0	1.5-2.5	40-60	30-60
H. Alpha-olefins C range: C_{12} to C_{18} single or mixture from Shell (Shop route)	1.5-2.5	1.0-2.0	40-60	40-70

* Feedstock comments see next page

** 5% AD solution, 40 mm cell N42 filter, unbleached, unless specified otherwise.

Ballestra assumes organic feedstock specifications as outlined in chapter 3, raw materials:

For linear alkylbenzene see 3.6.1. alkylbenzene quality;

For primary alcohols see 3.7.1., primary alcohol quality;

For primary alcohol ethoxylates see 3.8.1., ethoxylate quality;

For alpha olefins see 3.9.1., alpha olefin quality;

For fatty acid me-ester see 3.10.1., me-ester quality;

For caustic soda see 3.11.1., caustic soda quality.

The free oil guaranteed figures are achievable when the organic raw material sulph(on)atability is 99.0% (the sulphonatability would be established by analysis of the paraffin content in the free oil extracted from the acid after the sulphonability test) and with a 99.5% yield of the sulphonable compound.
Example for linear alkylbenzene (LAB) molecular weight LAB = 245.
Molecular weight of linear alkylbenzene sulphonate (LABS) = 245 + 80 + 22 = 347.
Sulphonability (S) 99.0%;
Sulphonation yield (Y) 99.5%;
Free oil on 100% AM (AD):

{100 - SY/100} x {M.W. LAB/M.W. LABS} x {10000/SY} =

1.495 x 0.70605 x 1.0152 =

1.07% free oil on 100% active matter.

Table 22 Final product specification for Chemithon falling-film reactor (FFR)

FEEDSTOCK*	free oil on 100% AD	sulphate on 100% AD	Klett** colour	AD conc. in paste %
A. Branched DDB i.e. Oronite 60 (tri-decylbenzene C_{13})	1.5	1.5	acid: 100 paste:80	40 min.
B. Linear Alkylbenzene Witco Dodane S	1.1	1.3	26	60-70
C. Natural lauryl alcohol C_{12}-C_{14}	1.3	1.5	3	30-75
D. Synthetic alcohol C_{12}-C_{14} Ethyl 12-14	1.5	1.5	3	30-75
E. Ethoxylated alcohol Shell Neodol 25-3EO	2.0	0.5	10	27 (60-70)
F. Ethoxylated alcohol Conoco Alfol 12-14, 2EO	2.0	0.5	5-10	27 (60-70)
G. Tallow alcohol P & G	4.0	2.0	60	30-60
H. Alpha-olefin Shell Shop C_{14}-C_{16}	2.0	2.0	40	40-70

* Chemithon supplied the following feedstock specifications:
Oronite 60 and Nalkylene 500: Bromine No. 0.02 max.; dialkyl content 0.5 max.; moisture clear at 70 °F; unsulphonatable 1.3% max..
Natural lauryl alcohol; Iodine value 0.2 max.; moisture 0.1% max.; unsulphonatables 0.5% max..

** Klett 5% AD solution, 40 mm cell, N42 filter

Table 23 Final product specification for Chemithon Jet Impact Reactor

FEEDSTOCK* **	free oil on 100% AD	sulphate on 100% AD	Klett*** colour	AD conc. in paste %
A. Branched DDB i.e. Oronite 60 (tri-decylbenzene C_{13})	1.5	2.0	acid:150 paste:150	40
B. Linear DDB type Nalkylene 500	1.5	2.5	acid: 40 paste:40	60-70
C. Natural lauryl alcohol C_{12}-C_{14}	2.5	0.5	30	30-75
D. Synthetic alcohol C_{12}-C_{14} Ethyl C_{12}-C_{14}	3.0	0.5	40	30-75
E. Ethoxylated nat. lauryl alcohol with 3 EO	3.0	0.5	40	27 60-70
F. Ethoxylated C_{12}-C_{14} - 2 EO, e.g. Conoco Alfol 12-15, 2 EO	2.5	0.5	30	27 60-70
G. Alpha-olefin Shell C_{12}-C_{14}	3	2	60 (bleached)	40-70

* Chemithon supplied the following feedstock specifications: Oronite 60 and Nalkylene 500: Bromine No. 0.02 max.; dialkyl content. 0.5 max.; moisture clear at 70°F; unsulphonatables 1.3% max. Natural lauryl alcohol: Iodine Value 0.2 max.; moisture 0.1 max.; unsulphonatables 0.5% max.

** Natural and synthetic fatty alcohols - no recent data

*** Klett 5% AD solution, 40 mm cell, N42 filter

Table 24 Final product specification for Meccaniche Moderne falling-film reactor

FEEDSTOCK	free oil on 100% AD	sulphate on 100% AD	Klett* colour	AD conc. in past %
A. Branched DDB Rhone-Poulenc	2.2	1.6	acid: 80 paste:80	65 max.
B. Linear DDB LAB 243 Shell	1.0	0.8	acid: 25 paste:25	70 max.
C. Natural lauryl alcohol C_{12}-C_{14} Lorol special C_{12}-C_{14}	1.5	0.8	20	35
D. Natural lauryl alcohol C_{12}-C_{18} Technical Lorol C_{12}-C_{18}	2.0	1.0	40	30
E. Synthetic alcohol C_{12}-C_{14} Alfol C_{12}-C_{14} Condea	1.8	1.8	20	30
F. Ethoxylated alcohol C_{12}-C_{14} nat.				
Denidol C_{12}-C_{14} 2/3 EO	1.5	0.8	30	30 and 65/70
Alfol Condea C_{12}-C_{14} 2/3 EO	1.8	0.8	40	30 and 65/70
G. Natural fatty alcohol type TA 1618 (P&G)	4.0	1.2	60	28
H. Synthetic fatty alcohol C_{16}-C_{18} Alfol Condea C_{16}-C_{18}	4.0	1.2	60	27
I. Alpha-olefin C_{12}-C_{14} Shell C_{12}-C_{14} linear	2.0	1.2	60	40
J. Alpha-olefin C_{14}-C_{16} Shell C_{14}-C_{16} linear	2.5	1.3	80	40

* Klett 5% AD solution, 40 mm cell, N42 filter

Table 25 Final product specification for Mazzoni multi-tube falling-film reactor (MT-FFR)

FEEDSTOCK	free oil on 100% AD	sulphate on 100% AD	Klett* colour	AD conc. in paste %
A. Linear alkyl-sulphonic benzene	1.0	1.0	50	acid: 97.6
B. Linear alkyl-benzene	1.0	1.0	20	45
C. Synthetic alcohol C_{12}-C_{14}	1.0	1.2	10	30
D. Natural alcohol C_{12}-C_{14}	1.2	1.4	20	70
E. Natural or Ziegler alcohol ethoxylated C_{12}-C_{14} - 2 or 3 EO	1.0 1.3	1.1 1.3	15 20	30 75
F. Oxo alcohol C_{12}-C_{14} ethoxylated with 2 or 3 EO	1.5 2.0	1.2 1.4	30 40	27 70
G. Synthetic alcohol C_{16}-C_{18}	2.6	1.4	40	35
H. Alpha olefin C_{14}-C_{16}	1.8	1.4	50	40
I. Fatty acid Methyl Ester C_{16}-C_{18} or mixture C_{12}-C_{14}	1-2	1.5-2.5	80-100**	30-40

* Klett 5% AD solution, 40 mm cell, N42 filter

** after bleaching

Equipment suppliers' guaranteed final product specification

Table 26 Final product specification for Lion Corp T-O FFR

Raw Materials		Products (Na salt)			
species	suppliers	unreacted* oil %	Klett col.**	Na_2SO4 %	A.M. concentration %
Linear Dodecil Benzene	Mitsubishi Petrochemical Nippon Petroleum Detergents	1.0	17	2.0	50
C_{10}-C_{14} Alcohol	New Japan Chemical	2.3	20	1.3	35
C_{12}-C_{15} Alcohol	Mitsubishi Petrochemical	1.9	28	1.5	30
C_{12}-C_{13} Alcohol Ethoxylate (p=2)	Mitsubishi Petrochemical & Lion Chemical	1.5	22	0.6	30
C_{12}-C_{13} Alcohol Ethoxylate (p=3)		1.0	15	0.4	70
C_{12}-C_{15} Alcohol Ethoxylate (p=3)		1.4	14	0.5	
C_{14} Alpha Olefin	Mitsubishi Chemical Gulf Oil Chemical Shell Chemical	2.3	35	2.0	45
C_{14}-C_{18} Alpha Olefin		2.1	45	2.8	40
C_{16}-C_{18} Alpha Olefin		2.5	50	3.1	40

* based on A.M. (Active Matter)
** 5% A.M. solution
40 mm cell blue filter No. 42

Table 27 1,4-dioxane* and free oil levels indicated by various suppliers (ppm (mg/kg) dioxane and % free oil on 100% AD)

FEEDSTOCK**	Ballestra		Chemithon		Mazzoni		Mec.Mod.	
	dioxane	free oil	dioxane	free oil	dioxane	free oil	dioxane	free oil
Natural alcohol (type Dehidol LS-2) or synthetic (type Dolocenol 25/2)	10-30	2.0	10-30	2.0	40	2.0	40-50	2.0
same with 3EO	10-30	2.0	10-30	2.0	40	2.0	40-50	2.0

* Even though the above values represent the guaranteed concentration of dioxane contamination in ether sulphates from the suppliers, in practice the formation of dioxane is extremely dependent upon the overall quality of operation of the sulphonation plant. Under ideal conditions the performance of the Ballestra, Chemithon, MM and Mazzoni plants is unlikely to be significantly different.

** Assuming alcohol ethyloxylate specifications as outlined in section 3.8.1.

6.2 In-plant analytical methods to check product quality

The contractors' guaranteed product specifications for the various feedstocks are summarised in the previous section. The recommended test procedures for LABSA, LABS and LES are given below, table 28.

Table 28 Analytical test schedule for LABSA, LABS, LES

Test	Apparatus*	Time** (in min.)	Frequency
Alkylbenzene sulphonic acid (LABSA)			
post-sulph. reactor			
Acid value	GLE	5	2/hour
post-stabiliser			
Acid value	GLE	5	2/hour
Water	Karl Fisher	10(5)	4/shift
Colour	Klett	5	4/shift
Free oil	GLE	240	1/shift
Alkylbenzene sulphonate paste (LABS)			
AM	GLE	15	2/shift
free oil	GLE	240(45)	1/shift
Colour	Klett	5	2/shift
pH	pH-meter	5	2/shift
Lauryl ether sulphate paste (LES)			
AM	Hyamine titration	15(5)	2/shift
pH	pH-meter	5	2/shift
Colour	Klett	5	2/shift
Free oil	HLPC	15	1/shift
dioxane	GC(H/space)***	15(5)	1/hour

* GLE - General Laboratory Equipment
** Actual time of operator direct involvement shown in bracket
*** Gas Chromatography Head Space

Detailed analytical method references are given in appendix 4. The data were supplied by Ballestra SPA - Milano.

6.3 Linear alkylbenzene sulphonic acid (LABSA) quality, storage, handling and safety

6.3.1 LABSA quality

Factors which affect the quality are summarised below:

i)	Organic feedstock quality bromine value, moisture and sulphonatability	See 3.6.1. alkyl-benzene quality
ii)	Sulphonation conditions like mole ratio and temperature	See 4.2.1., table 5 sulphonation chemistry

For heavy duty liquid products the maximum sulphuric acid levels required may be 1% or even lower (0.6%). The sulphuric acid level will reduce on prolonged ageing (before hydrolysis) due to the equilibrium between sulphuric acid, anhydride and pyrosulphonic acid (see 4.2.1., alkylbenzene sulphonation chemistry). Consequently, increased ageing times and increased temperatures should reduce the sulphuric acid and free oil at the expense of somewhat poorer colours. The ageing could take place in tanks in a batch-wise operation or in a continuous tank reactor, approaching plug flow characteristics. After prolonged storage, hydrolysis should take place as described in 5.5.10.

6.3.2 LABSA safety information

LAB sulphonic acids are eye and skin irritants. When handling sulphonic acids, safety goggles/face shields, impermeable gloves and a protective apron should be worn. A supply of clean, running water should be available. In case of accidental contact, wash off splashes on the skin using copious supplies of water. With accidental eye contact, wash out the eye with copious supplies of clean water and get medical attention.

6.3.3 Storage and handling

The corrosive behaviour of the anhydrous sulphonic acid is similar to that of concentrated sulphuric acid (100%). It can be stored in mild steel vessels as long as there is no moisture ingress. The use of mild steel tanks can result in a small pick-up of iron which can cause discolouration of the final product; such tanks are therefore not preferred. Consequently, the stainless steel 304 and 316 types are widely used for the storage of sulphonic acids.

Where stainless steel is not available, mild steel tanks, preferably coated with a suitable stoving lacquer, e.g. epoxy phenolic stoved lacquer, may be used for storage.

Since sulphonic acid is a highly viscous material at ambient temperatures, it is recommended that, to facilitate handling (e.g. pumping from bulk storage to neutralisation), it is maintained at about 30-35°C for a time span not longer than 2 months. The recommended heating facility for the storage tank is a warm water circulation system.

There are no problems with flow, pumping, pipe cleaning, etc. At temperatures between 40°C and 45°C, sulphonic acids can be handled with ordinary pumps (centrifugal, positive displacement) in pipes of normal diameters and lengths. The flow will always be laminar at the velocities applied (0.5 m/s -1.0 m/s) and pressure drops can be calculated from the Poiseuille equation:

$$\Delta P = 32.\eta Lv/D^2 \text{ in which}$$

ΔP =	pressure drop	N/m^2
η =	dynamic viscosity	Ns/m^2
L =	pipe length	m
v =	velocity	m/s
D =	pipe diameter	m

Experience in handling the acid has shown that the lines should be stainless steel and that, outside buildings in cold climates, they should be hot water trace-heated and lagged. The recommended pump is a positive displacement type with stainless steel valves with PTFE sealing.

6.3.4 ENGINEERING DATA - PHYSICAL PROPERTY DATA SHEET

NAME:	(LINEAR) ALKYLBENZENE SULPHONIC ACID (LABSA)				
CHEMICAL FORMULA:	$C_{18}H_{29}SO_3H$,		MW 326 (range)		
APPEARANCE:	Amber viscous liquid				
DENSITY:	LINEAR	kg/m^3	1050	1045	1040
		°C	30	40	50
	BRANCHED:	kg/m^3	1075	1070	1060
		°C	30	40	50
VISCOSITY: Ns/m^2	Marlon AS 3		Dobane 103		Oronite 60 (branched)
25°C	1.6		1.6		20
40°C	0.6		0.6		3
60°C	0.2		0.2		0.6
SPECIFIC HEAT:	kJ/kg°C		1.6		
HEAT CONDUCTIVITY:	W/m°C		0.13		

6.4 Linear alkylbenzene sulphonate paste (LABS) quality, storage, handling and safety

6.4.1 LAB sulphonate paste quality

Factors which affect quality are summarised in table 29.

Table 29 Factors which affect LABS quality

i)	Organic feedstock quality: bromine value, moisture and sulphonatability	See 3.6.1. alkylbenzene quality
ii)	Neutralising agent (caustic soda) quality notably iron contamination	See 3.11.1. caustic soda quality
iii)	Sulphonation and neutralisation conditions	See 4.2.1., table 5 and 4.3.2. LABSA neutralisation, and 5.6.1. Chem. neutralisation, and Engineering aspects of neutralisation

The AD (active detergent) concentration in pastes normally varies between 40 and 60%, depending on rheological properties of the neutralised sulphonic acid type. As LAB sulphonate paste contains substantial amounts of water and is difficult to store and handle the stable LAB sulphonic acid is preferred as the intermediate product for storage, transport and handling, followed by neutralisation in-situ in the actual final detergent product manufacture.

6.4.2 LAB sulphonate safety information

As neutral aqueous salts, these materials are relatively innocuous. Like all concentrated detergents, they can cause eye irritation and repeated or prolonged contact can cause skin irritation. Eye protection should be worn if there is a recognised risk of splashing. Repeated or prolonged skin contact should be avoided by using impermeable gloves and other appropriate clothing. Wash exposed skin as soon as possible after contamination. If eye contamination occurs, wash the eye with running water for 15 minutes.

6.4.3 Storage and handling

These so-called pastes, with an active detergent content between 40 and 60%, are viscous liquids (see 5.6.1.) which also show a distinct tendency to separate into two phases - an upper active phase and a lower aqueous phase. The separation tendency, unless countered, will result in serious variations in composition throughout the mass of the paste, and these will be reflected in variations in the active content of products, for example NSD powders or liquids.

Due to the high viscosities of the pseudoplastic paste, the separation tendency and the diluted form of AD pastes, sulphonic acid storage is preferred to paste storage. The actual neutralisation is preferably carried out in-situ i.e. as part of the detergent product manufacturing process.

When in-situ neutralisation is not possible, pastes are stored in agitated tanks of up to 100 tons capacity. Heating coils should be provided with hot water (80 °C). Pitting of mild steel will eventually occur and, as freedom from iron contamination is important for on paste colour, stainless steel or glass-reinforced polyester tanks should be used.

For final products that require alkaline silicate, silicated paste can be stored. For this type of paste with a composition of 34% AD, 2% salts, 14% alkaline silicate and 50% water, mild steel tanks can be used as alkaline silicate is a recognised mild steel corrosion inhibitor. Its inclusion in alkylbenzene sulphonate paste, besides aiding homogeneity, enables the paste to be handled satisfactorily in mild steel equipment without becoming contaminated with iron. Agitation in silicated paste is not required; circulation with an external pump, at a rate sufficient to turn over the contents once per hour, will be adequate.

Mono-type pumps are recommended for alkylbenzene sulphonate paste duty. Pipe diameter selection has been discussed in section 5.6.1..

6.4.4 ENGINEERING DATA - PHYSICAL PROPERTY DATA SHEET

NAME:	(LINEAR) ALKYLBENZENE SULPHONATE PASTE					
CHEMICAL FORMULA:	$C_{18}H_{29}SO_3Na$ MW 348 (range)					
APPEARANCE:	White opaque highly viscous liquid					
DENSITY:	kg/m^3	variable with AD concentration an appropriate figure 1000				
VISCOSITY:	Ns/m^2 at a shear rate of 30 s^{-1}					
	40 °C	Dob 055	(40% AD)	2.0 Ns/m^2		
		Petr. Kor	(42% AD)	2.2 Ns/m^2		
		Dob 102	(55% AD)	3.5 Ns/m^2		
	60 °C	Dob 055	(40% AD)	1.2 Ns/m^2		
		Petr. Kor	(42% AD)	1.0 Ns/m^2		
		Dob 102	(55% AD)	1.6 Ns/m^2		
SPECIFIC HEAT:	kJ/kg °C					
AD concentration	%	35	40	45	50	
Specific heat		3.3	3.2	3.1	3.0	
HEAT CONDUCTIVITY:	W/m °C					
AD concentration	%	35	40	45	50	
Heat conductivity		0.43	0.40	0.38	0.35	

6.5 Primary alcohol sulphates (PAS)

6.5.1 Primary alcohol sulphate quality

Factors which affect the quality are summarised in table 30.

Table 30: Factors which affect PAS quality

i)	Organic feedstock quality	See 3.7.1. primary alcohol quality
ii)	Neutralising agent quality	See 3.11.1. caustic soda quality
iii)	Sulphonation and neutralisation conditions	See 4.2.2. table 5 sulphonation chemistry, 4.3.3. neutralisation, 5.6.1. Chem.Engineering aspects neutralisation

Paste concentrations are 30% AD/75% AD for Coco-type and 30% AD/60% AD for tallow-type primary alcohol sulphates.

6.5.2 Primary alcohol sulphate safety information

See 6.6.2.,alcohol ether sulphate safety information.

6.5.3 Storage and handling of PAS

The primary alcohol sulphate pastes with an active detergent varying between 30% and 75% AD are viscous liquids (see 5.6.1. and 6.6.3.)The apparent viscosity is dependant on AD concentration, temperature and the concentration of sodium salts (NaOH, Na_2SO_4). High AD pastes tend not to deaerate and may therefore cause problems in bulk transfer. Pastes are stored in mildly agitated stainless steel vessels. Solidification should be avoided and therefore heating coils with hot water are required. The hot water temperature should be 45°C for Coco-PAS and 60°C for Alfol (C16 - C18) PAS.

During storage, solutions of alcohol sulphates often tend to become acidic due to hydrolysis. The rate of hydrolysis depends on the storage conditions (pH, T) but also on the severity of the sulphonation and neutralisation conditions. Sulphates (PAS) are stable under slightly alkaline conditions (pH about 8.5 and higher). Decomposition is slow at ambient temperatures (20°C), but may become rapid enough to cause appreciable degradation if the product is exposed to temperatures of 60°C or higher for more than a few hours and the pH is below 8-9. Further, as sulphuric acid is the product of hydrolysis (see 4.3.3.), the process is auto-catalytic and, if unchecked, can lead to spoilage of the product. As an aid to reaching pH values between 8 and

10 during and after neutralisation the normal practice is to buffer the product, e.g. with a mixture of a 10% solution of carbonate/bicarbonate or citrate/phosphate. A level of 0.5 - 1% on active matter is suitable.

The pH of the buffering system mixture carbonate/bicarbonate for example can be calculated as follows:

$HCO_3^- + H_2O \rightleftharpoons CO_3^{2-} + H_3O^+$

$[CO_3^{2-}][H_3O^+] / [HCO_3^-] = K_a$

K_a is the dissociation constant of the HCO_3^- acid with a value of $5x10^{-11}$ (P_k = 10.25)

$[H_3O^+] = K_a \times [HCO_3^-] / [CO_3^{2-}]$

With equal amounts of $NaHCO_3$ and Na_2CO_3 in the buffering solution the pH will be equal to pK_a i.e. 10.25.
For some product applications a lower pH is required and different buffering systems will be used. (0.1-0.2% on 100% active detergent of citric acid or phosphoric acid to be added after neutralisation, resulting in a final pH of 7-8).

Mono-type pumps, lobe pumps or gear pumps (viking) are recommended for PAS paste duty. Pipe diameter selection has been discussed in section 5.6.1., chemical engineering aspects related to neutralisation reactor systems.

6.5.4 ENGINEERING DATA - PHYSICAL PROPERTY DATA SHEET

NAME:	PRIMARY ALCOHOL SULPHATE			
CHEMICAL FORMULA:	$C_nH_{2n+1}OSO_3Na$			
APPEARANCE:	White opaque highly viscous liquid			
DENSITY:	kg/m^3	Temp. 20°C	1040	
VISCOSITY: 20°C	Ns/m^2	C_{12}-C_{13} 29% AM 0.3-0.4	C_{12}-C_{15} 28% AM 2-3	C_{14}-C_{15} 20% AM 7-9
40°C	Ns/m^2	0.1-0.2	0.5	2-3

6.6 Primary alcohol ethoxysulphates, lauryl ether sulphates (LES) quality, storage, handling and safety

6.6.1 Factors which affect the quality are summarised in table 31.

Table 31 Factors which affect LES quality

i)	Organic feedstock quality	See 3.8.1. ethoxylated alcohol quality
ii)	Neutralising agent quality	See 3.11.1. caustic soda quality
iii)	Sulphonation and neutralisation conditions	See 4.2.3. ethoxylated alcohol sulphonation chemistry. Table 5, 4.3.3., 5.6.1. Chem. Engineering aspects related to neutralisation reactor systems

Control of the level of 1,4-dioxane in the ethoxylated alcohol sulphates is often seen as a good measure of the overall performance of the sulphonation plant. All equipment suppliers have examined the effects of various operating and design parameters on the formation of 1,4-dioxane.

The results of these studies have been published by Ballestra, MM and Chemithon.
Ballestra conclude that the 1,4-dioxane content increases dramatically with a slight excess of the SO_3/organic mole ratio in each reactor tube and any part of each reactor tube and in the overall process mass balance. Furthermore, 1,4-dioxane levels are affected by the type of base alcohol, the range of ethoxylation and the moisture content in the feedstock. The 1,4-dioxane content in alcohol ethoxylate sulphates in the C_{12}-C_{14} range with 2-3 EO molecules, can nowadays be specified at levels well below 50 ppm on 100% active matter.

Chemithon findings are in agreement with the above statements: mole ratio was found to be the most important variable. Mole ratios between 1.01 and 1.03 yielded 1,4-dioxane values from less than 20ppm to 50ppm on 100% active matter with Shell feedstock Neodol 25-3 (3 moles EO). Free oil concentrations were 2.0% on 100% active matter. Higher mole ratios caused a dramatic increase in 1,4-dioxane. Mole ratios above 1.05 easily led to levels of several hundred ppm. From these results, it is clear that precise raw material delivery systems must be employed.

For purposes of reducing 1,4-dioxane, control of the mole ratio was found to be a much broader concept than simply ratioing reactants to the reactor. The mole ratio had to be controlled at all points in the reactor. This means that the organic films in the reactor have to be controlled so that they are of equal thickness (mass) at each circumferential cross-section of the reactor. The gas flow has to be controlled so that it is uniform at all points around the reactor circumference. In addition, the organic film has to be introduced into the reactor so that organic mist does not form. To accomplish this, Chemithon developed a new precise organic distribution system to ensure liquid film uniformity ($\pm$0.5%) on the reactor surfaces. The reactor geometry provides the gas flow control required to ensure uniform SO_3 gas supply to all points in the reactor.

In a fixed geometry reactor, SO_3 gas concentration cannot be varied without changing either the reactor's organic liquid loading or the gas velocity (see also 4.3.3, chemical reaction engineering aspects related to sulph(on)ation in film reactors), both of which have an effect on 1,4-dioxane levels. In general, with organic liquid loadings (mass of feedstock per unit of reaction surface circumference) in the normal commercial ranges, SO_3 concentrations equal to or below 3% gave acceptable results while those of 4% or above did not. With an SO_3 concentration of 3%, the gas velocity must be increased to ensure an adequate SO_3 flux to the organic phase, required for complete conversion.

Higher gas velocities at a fixed SO_3 concentration in the Chemithon-design film reactors reduced 1,4-dioxane levels. The higher gas velocities achieved this result by reducing the reactor residence time of the organic phase in the falling-film part (residence times of 10 seconds or less) and by improving the heat and mass transfer rates in the reactor.

Residence time in the recycle quench loop of the Chemithon FFR quench loop system was found to have an influence on 1,4-dioxane levels. However, the effect was not pronounced if the quench loop temperature was maintained at the optimum value (not released in Chemithon publication).

The use of the post-reactor quench loop was found to be beneficial for low levels of 1,4-dioxane, provided that the residence time in this part of the reactor system was not longer than approximately five minutes. The recycle loop quench temperature had a strong effect on 1,4-dioxane levels, second only to that of mole ratio.

Finally table 32 indicates the optimal conditions for sulphation in film reactors and neutralisation of the lauryl-type ethoxylate.

Table 32 Ethoxylated alcohol sulphation and neutralisation - optimal conditions

1. Mole ratio SO_3: organic feedstock 1.01-1.02

2. Moderate effectively the temperature rise due to the exotherm to control dioxane formation.
    - SO_3 in air → 3.0% vol. max.
    - SO_3 in air temperature → 45 °C
    - Organic feedstock temperature → 30 °C
    - Cooling water upper section FFR → 30 °C
    - Throughput: 85% w/w of nominal for alkylbenzene.

3. Neutralise quickly to limit dioxane formation.

4. Avoid pH < 7. Acid conditions enhance dioxane formation and, additionally, hydrolysis will occur (autocatalytic reaction) with potentially disastrous consequences.

5. Ensure excellent mixing on micro-scale during neutralisation to avoid acid spots.

6. Cool 70% active matter ether sulphates from 50 °C (neutralisation production temperature) to 40 °C.

Figure 45 Correlation dioxane level vs. free oil

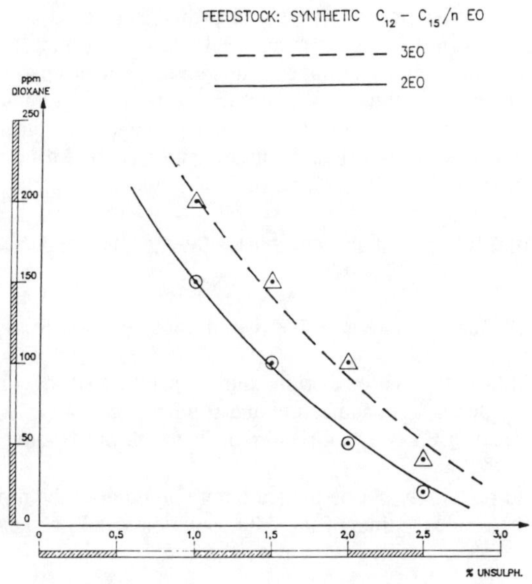

Figure 46 Influence of mole ratio SO_3/organic on dioxane

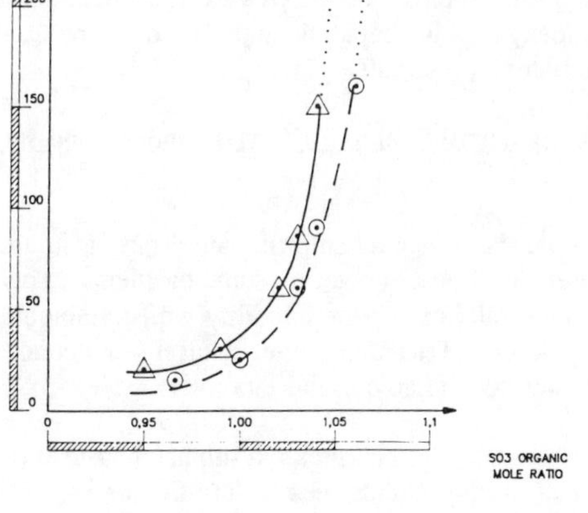

6.6.2 Alcohol ether sulphate safety information

The undiluted product can cause eye irritation and repeated or prolonged contact can cause skin irritation. Eye protection should be worn if there is a recognised risk of splashing. Repeated or prolonged skin contact should be avoided by using impervious gloves or other appropriate clothing. If eye contamination occurs, wash the eye with running water for 15 minutes. Wash exposed skin as soon as possible after contamination. Ether sulphates containing ethanol are classed as "flammable liquids" and consequently should comply with local and national regulations pertaining to their storage and handling (see also 3.17., ethanol).

6.6.3 Microbiological control of dilute alcohol ether sulphates (and alcohol sulphates, 27% AM)

Application of the following guidelines will minimise the risk of contamination from the source:

(i) Concentrated SLES (at 70% AM) is self-preserving and is therefore strongly preferred to 27% SLES. This applies to purchase, manufacture and storage of SLES.
Note that concentrated SLES does not require dosing with preservative.

(ii) Concentrated SLES should preferably not be diluted before addition to the process. If pre-dilution is unavoidable, the residence time of the dilute solution should be kept as short as possible.

(iii) All dilute SLES should contain preservatives, even if stored for only a short time. The recommended dosage is 0.06 -0.08% (medical) formaldehyde in the SLES solution, i.e. 0.2% formalin (40% formaldehyde strength) to 0.27% formalin (30%). If dilute SLES is bought in, the purchase specification should include a certificate of quality, including a guarantee of effective microbiological contamination control.

(iv) Regular microbiological checks should be carried out on the received and stored material.

(v) Other dilute active detergent raw materials should be similarly monitored for contamination and preservative dosing applied. In some cases (e.g. ammonium lauryl ether sulphate) however, formaldehyde is incompatible and should not be used. Myacide BT (0 - 1%) is an effective alternative preservative.

6.6.4 High-active alcohol ether sulphates (approx. 60 to 70% AD) handling and storage

Shampoo bases have traditionally been blends of alcohol sulphate and a fatty acid alkanolamide, the latter functioning as a foam stabiliser. There are, however, some problems associated with alcohol sulphates - for example the sodium salts have poor solubility whilst ammonium salts, although more soluble, are potentially odorous. Triethanol amine neutralised alcohol sulphates have good solubility characteristics but are poor in colour and more expensive.

The above factors have led to an increasing use of alcohol ethoxysulphates instead of alcohol sulphates in products like shampoos, dishwashing liquids, heavy duty liquids and bath products. The advantages of C_{12}-C_{15} with 2-3 EO per mole alcohol ethoxysulphates are:

- good solubility and colour;
- low irritancy potential;
- ease of viscosity adjustment.

As the alcohol ethoxy-sulphate acid is unstable, the product has to be stored and transported in the neutralised form. The "lauryl" ethoxy-sulphates are commercially available in concentrated forms - 60% AM with ethyl alcohol and 70% AM without any addition. The 70% high-active product is a fluid, pumpable material which is made without using viscosity modifiers such as ethanol.

As the concentration of an anionic surfactant in solution is increased from 20 to 70%, the viscosity of the solution generally increases, but may eventually undergo a sudden decrease. The grades of C_{12}-C_{15}, 2-3 EO alcohol ethoxy-sulphates which are commercially available, are normally in the concentration range of 25-30% AM or 55-70% AM. Between about 30% AM and 55% AM concentration (the exact figures depend on the specific product and the amounts of inorganic salts and free oil present), ethoxysulphates form a highly viscous "gel" which corresponds to the M (middle) phase (see fig. 47).

Between about 55% AM and 75% AM, they exist as a "pourable gel", which corresponds to the G or neat phase. This phase has a lamellar structure which allows the product to flow, increasingly so as greater shearing force is applied (see also 5.6.1.). This allows the material to be handled and pumped with relative ease. An alternative technique for obtaining a low-viscosity, high-active content ethoxysulphate is to add ethanol as a solubiliser. In this way a mobile liquid product can be obtained containing a nominal 60% active matter product and about 14% ethanol.

Figure 47 Phase structures of LES pastes

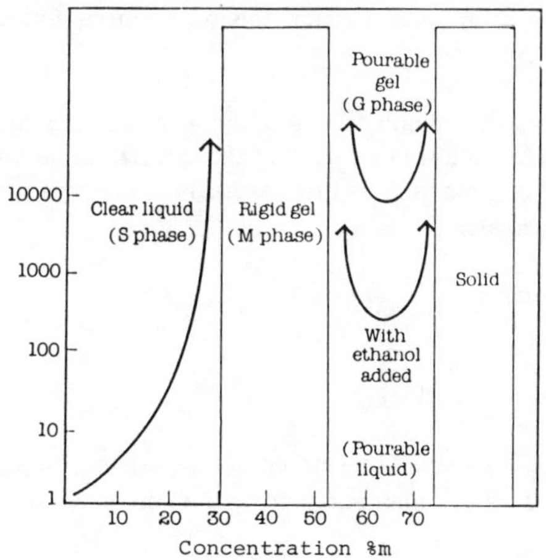

concentration percentage of mass

The presence of ethanol in an ethoxysulphate is acceptable if the ultimate end-product itself contains ethanol. For other applications such as toiletries, however, ethanol may be undesirable. Traditionally, such requirements were satisfied by the 25-30% AM products, but are now increasingly being met by nominal 70% AM products containing no added solubiliser. From the customer's point of view, such a product offers the advantage of lower transport and storage costs and also the opportunity to formulate a product of higher total active content than would otherwise be possible.

A less obvious advantage of a 70% AM ethoxysulphate is its much greater resistance to bacteriological contamination than a conventional 25-30% AM material.

For dilution of 70% AM product, the importance of following the correct procedure should be mentioned. As can be seen from figure 47, addition of water to a 70% AM ethoxysulphate will quickly reduce the active content to a point where the system adopts a middle (M) phase structure which is an almost-rigid gel, and very difficult to disperse in water. To avoid this problem, the high-active ethoxysulphate should be added slowly to water, preferably warm water (35-40°C) with continuous but not vigorous agitation. In this way, the mixture remains liquid and any particles of gel formed will soon dissolve. Similar remarks apply to the dilution of the 60% AM product, although here the presence of ethanol assists the dispersion of the gel.

To reach the 70% AM level during start-up of the neutralisation reactor in ether sulphate manufacture, it is necessary to follow a process route which deviates from the usual method of neutralisation. The normal technique of start-up is to fill the loop with water and caustic soda, then dose the organic acid to the loop. By following this procedure, the paste concentration increases gradually from zero to the steady-state concentration. When applying this route to lauryl ether sulphates (LES), the very viscous region between 30 and 60% AM has to be passed, causing a situation where the circulation pumps may not handle the extremely viscous paste. This will lead to an uncontrolled situation in the neutraliser. With very precise control of excess caustic and extra design features in the neutralisation plant, this route can be followed during start-up of 70% SLES paste manufacture.

An alternative is to avoid the highly viscous region by filling the neutralisation equipment from the very beginning with 70% LES paste, followed by starting the circulation pump and the in-line homogeniser and subsequently dosing the various raw materials at the right set-points so that a 70% concentration is produced immediately.

The recommended processing conditions are:

- Paste loop temperature 40-45°C
- Cooling water temperature 30-35°C

70% SLES (sodium "lauryl" ether sulphate) should be stored at temperatures between 15-40°C since below 10°C the material is too thick to pump and above 40°C there is a danger of hydrolysis.

Aeration can be a major problem for pumping 70% SLES. Aeration occurs during sulphonation. Deaeration before neutralisation may lead to an unacceptable increase of 1,4-dioxane due to extra residence time of the acid product in the deaeration system. Therefore deaeration of the paste prior to storage may be preferred.

Pastes are stored in mildly agitated stainless steel (304 L, 316 L) tanks. Polyester resins form suitable tank coatings when only mild steel tanks are available. Copper and its alloys may cause discolouration of the product and should be avoided. Temperatures below 15°C should be avoided and therefore heating coils with warm water (40°C) are required. During storage, solutions of alcohol ether sulphate often tend to become acidic due to hydrolysis. Ways to prevent this have been described in section 6.5.3. and 6.6.4., storage and handling of primary alcohol sulphates.

Positive displacement pumps like for example mono-type pumps are recommended for all ether sulphate paste duties. Pipe diameter selection has been discussed in section 5.6.1., chemical engineering aspects related to neutralisation reactor systems. Paste velocities of approximately 0.5 m/s should be the basis for pipe diameter calculations.

Transport of 70% ether sulphates is often carried out in drums. The drums should have removable polyethylene liners to ensure product quality. Care must also be exercised not to overheat the drums, because of potential damage to the polyethylene drum liners. Temperatures in excess of 50°C can cause the ethoxysulphate to hydrolyse and become corrosive. Transport in tankers can be considered but it is preferable to deaerate the product before tanker loading.

6.6.5 ENGINEERING DATA - PHYSICAL PROPERTY DATA SHEET

NAME: PRIMARY ALCOHOL ETHOXYSULPHATE

CHEMICAL FORMULA: $C_nH_{2n+1}O(CH_2CH_2O)_nSO_3Na$

1. DOBANOL 23 - 2 EO	27% AM (approx.)			
Mean molecular mass:	384			
Appearance at 20°C:	clear liquid			
Density at 20°C kg/m³:	1050			
Viscosity Ns/m²:	0.5			
Preservative:	may contain a small quantity of formalin to inhibit the growth of bacteria in the product			
Buffer:	contains a small quantity of sodium carbonate to stabilise pH			
2. DOBANOL 23 - 2 EO	70% AM (approx.)			
Mean molecular mass:	384			
Appearance at 20°C:	clear pourable gelled liquid			
Density at 20°C kg/m³:	1080			
Apparent viscosity Ns/m²:	7	3	1.2	
Shear rate s⁻¹:	10	50	100	
3. DOBANOL 25 - 3 EO	70% AM (approx.)			
Mean molecular mass:	441			
Appearance at 20°C:	clear pourable gelled liquid			
Density at 20°C kg/m³:	1080			
Apparent viscosity Ns/m²:	5		0.9	
Shear rate s⁻¹:	10		100	

6.7 Alpha-olefin sulphonates (AOS) quality

Factors which affect the quality of the neutralised hydrolysed product are summarised in table 33.

Table 33 Factors which affect AOS quality

i)	Organic feedstock quality	See 3.9.1. alpha-olefin quality
ii)	Neutralising agent quality	See 3.11.1. caustic soda quality
iii)	Sulphonation chemistry	See 4.2.4. alpha-olefin sulphonation chemistry
iv)	Neutralisation and hydrolysis	See 4.3.4. neutralisation/ hydrolysis of alkene sulphonic acids and sultones
v)	Bleaching	See 4.4. bleaching chemistry

In section 3.9. a general product specification for alpha-olefins is given. A more detailed specification is depicted in table 34.

Table 34 Specification of alpha-olefin sulphonates based on top quality C_{14}-C_{16} and C_{16}-C_{18} alpha-olefins

Active content	35-40%
Conversion	97% min.
Active composition alkene sulphonate	60-70%
3- and 4-hydroxyalkane sulphonates	20-30%
2-hydroxyalkane sulphonates	Nil
Disulphonates	5-10%
Free oil (on 100% AD)	2.5% max.
Na_2SO_4 (on 100% AD)	1.5% max.
NaCl	Nil
NaOH	0.1
Gamma sultones ppm	Not detectable
Delta sultones ppm	<10
Colour (Klett 5% AM)	80 max. (for C_{14}-C_{16}) 100 max. (for C_{16}-C_{18})

6.8 **Fatty acid methyl ester sulphonates quality**

Factors which affect the quality of the FAME sulphonates are summarised in table 35.

Table 35 Factors which affect FAME sulphonate quality

i)	Organic feedstock quality	See 3.10.1. FAME quality
ii)	Neutralising agent quality	See 3.11.1. caustic soda quality
iii)	Sulphonation chemistry	See 4.2.5. FAME sulphonation chemistry
iv)	Neutralisation chemistry	See 4.3.5. FAME acid neutralisation
v)	Bleaching chemistry	See 4.3.5. and 4.4. bleaching chemistry

6.9. **Bibliography**

1. Moretti G.F. Ballestra SpA, Milano. Private Communications.

2. Foster N.C. Chemithon, Seattle, USA, Private Communications.

3. Technical Brochures and Technical Proposals from Ballestra, Chemithon, Mazzoni, Meccaniche Moderne, Lion Co.

4. Huls, W-Germany. Marlon AS-3 Alkylbenzene sulphonic acid leaflet.

5. Unger Fabrikken A.S., leaflets for various qualities sulphonic acids.

6. Albright & Wilson Ltd., UK. Brochure: "High Active Surfactants".

7. Shell Detergent Range. (Shell Chemicals):
DOBANIC Acids (linear alkylbenzene sulphonic acids): "Properties and Handling". Technical Bulletin UD/O1O (1985).
DOBANOL 25-3S/27 (C_{12}-C_{15},3 EO, Sodium salt, 27% active matter), Data Sheet DI 2.5.2 (1979).
DOBANOL 25-2S/7O ,
DOBANOL 25-3S/7O, data sheets 2A 166 & 2A 165. DOBANOL 25-3A/6O (C_{12}-C_{15},3EO, Ammonium cation, 60% active matter, containing 13-15% mass ethanol).

8. Moretti G.F., Adami I. (1988). Ballestra SpA, Milano. "Product quality and reaction control for low dioxane content". CESIO 1988 paper, Paris.

9. Foster N.C. (1988). The Chemithon Corporation, Seattle, USA. "Application of pilot studies to minimizing 1,4 dioxane production during sulfation of ethoxylated alcohols".

10. Dahl B. (1989). "Processing, Storage and Handling of Lauryl Ether Sulphates". Paper presented at Practical Sulphonation Seminar, 18-20 April 1989, The Hague, The Netherlands. Center for Professional Advancement, P.O. Box H, East Brunswick, New Jersey 08816, USA.

11. Shell Detergents. "Preservation of liquid detergents and toiletries against bacterial spillage". Technical Bulletin UD/O17 (1984).

7 EXHAUST GAS AND EFFLUENT TREATMENT SO_3/AIR SULPHONATION PLANTS

7.1 Introduction

The need to meet local and national demands to restrict effluent levels from sulphonation plants makes it increasingly important that installed treatment systems are effective. The existing regulations around the world for both gaseous and liquid effluents vary from the non-existent to the stringent. Effluent carried by the exhaust gas leaving the sulphonation system consists of:

- SO_2, the sulphur dioxide leaving the sulphonation reactor system due to the incomplete conversion of SO_2 to SO_3. A small proportion of the SO_2 may arise from sulphonation side reactions;

- SO_3, SO_3/H_2SO_4 mist. The acid mists are formed as a result of residual moisture in the dried process air;

- organic acid mist which comes directly from entrainment from the sulphonation reactor system.

Three liquid effluent streams can be distinguished:

- the sulphuric acid collected at the bottom of the SO_3/air cooler system and in the SO_3 mist eliminator before the sulphonation reactor system;

- an approximately 10% solution of sodium sulphite/sulphate leaving the caustic scrubber;

- the dark coloured viscous liquid consisting of organic acid, sulphuric acid and 1,4-dioxane (when producing alcohol ethoxylate acids) dripping from the bottom of the electrostatic precipitator.

7.2 Treatment of the exhaust gas (see 5.9.)

7.2.1 Removal of residual SO_2 in exhaust gas leaving sulphonation system

SO_2 concentrations in exhaust gas leaving the sulphonation system will be, during steady-state operation, of the order of 2 000 mg/m^3, assuming 98% conversion in the SO_2-SO_3 converter tower. However, during plant start-up, concentrations of the order of 50 000 mg/m^3 may occur (see 5.9.2.). The level should be reduced to less than 5 mg/m^3 SO_2 in the exhaust gas leaving the caustic soda scrubber under steady-state conditions and about 15 mg/m^3 during initial start-up conditions.

7.2.2 Sulphuric acid mist removal from exhaust gas

Sulphuric acid mist levels are directly related to the dew-point of the dried air (see 5.2.2.). The sulphuric acid mist is effectively removed in the ESP and levels smaller than 10 mg/m^3 are guaranteed by ESP suppliers (see also 5.9.1. and 5.9.3.). Very small amounts of gaseous SO_3 will pass the ESP but will be trapped in the caustic soda scrubber.

7.2.3 Organic mist removal from exhaust gas

The amount of organic mist entrained in the exhaust gas leaving the sulphonation reactor system depends mainly on the organic feedstock quality. For example, ethoxy alcohol sulphate acids will easily form foams in the gas/organic liquid separator of the sulphonation system. Sulphonation equipment suppliers suggest the following levels of entrained organic material, based on weight percentage of organic feedstock:

- alkylbenzene 0.1-0.2%
- ethoxylated alcohol 0.2-0.3%

The organic mist is effectively trapped in the ESP (see also 5.9.1. and 5.9.3.), to levels not higher than 20 mg/m^3.

7.2.4 Chemical analysis of exhaust gas leaving the sulphonation plant stack (see fig. 48)

In order to obtain representative samples for analysis, a suitably constructed sampling point must be used. This normally takes the form of a ¼" outside diameter stainless steel pipe inserted into the centre of the duct leading to the stack and bent to face into the gas flow. This pipe is connected to the rest of the sampling train by as short a piece of silicone tubing as possible.

The sampling train consists of a plastic filter holder fitted with a 47mm Whatman GF/A glass fibre filter paper to trap sulphuric acid mist and organic acid mist, followed by three bubblers connected in series, each one fitted with a number 2 porosity sintered glass globe on the inlet pipe.

The first two bubblers are filled with 80/20 isopropanol/ water, with glass wool plugs wetted with the solvent to aid dissolution of the SO_2 and SO_3. The third bubbler is filled with 3% hydrogen peroxide as a back-up absorption system. The sampling pump used is a positive displacement type, fitted with either a flowmeter or a flow counter.

Figure 48 Chemical analysis of exhaust gas leaving the sulphonation plant stack

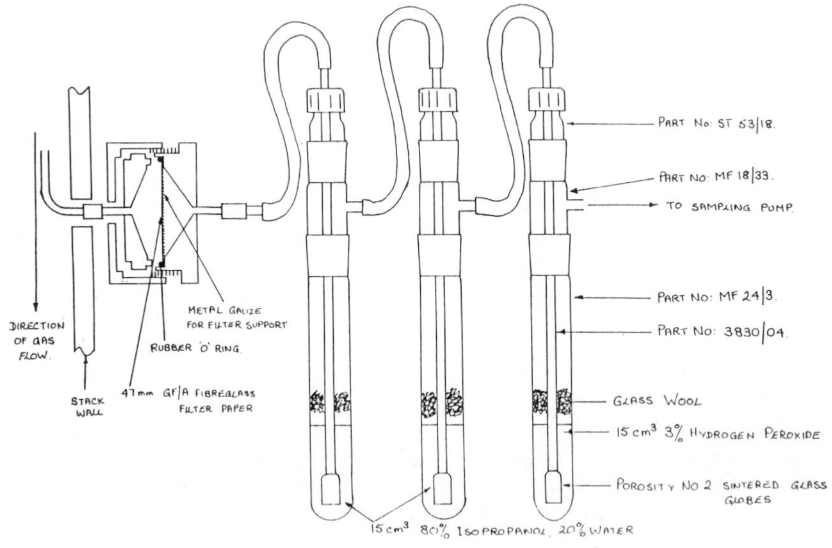

7.2.5 Summary table of various levels of gaseous effluents

Table 36 Gaseous effluent quantities mg/m³ before and after exhaust gas treatment system

	Before mg/m³	After mg/m³
SO_2 stationary conditions	appr. 2000	max. 5 (1 normal)
SO_2 start up conditions	appr. 5000	max. 15 (5 normal)
Sulphuric acid mist)	500-1000	max. 10
Organic acid mist)		max. 20

7.3 Treatment of liquid effluents

7.3.1 Treatment of collected oleum from SO_3 coolers, demisters and cyclones

The amount of sulphuric acid (oleum ±20%) collected in the bottoms of the SO_3/air cooler system and in the SO_3 mist eliminator, is directly related to the dewpoint of the dried process air (see 5.2.2.). With an air dewpoint of -60°C, only 2 kg oleum per 24 hours will be formed for a 1 ton AD/h sulphonation plant. With increasing dewpoint the amount of oleum collected will rise sharply. For sulphonation plants working with an SO_3 absorber (see section 5.2.6) it is recommended that the collected amounts of sulphuric acid/oleum be discharged into the vessel at the SO_3 absorbing system. The acid coming from the various parts of the gas raising plant should be collected via gravity flow in an intermediate tank with level control from where the acid is pumped off with a membrane pump to the SO_3 absorber. The use of compressed air is not recommended for safety reasons.

For sulphonation plants without an SO_3 absorber the following system is recommended: via gravity flow, sulphuric acid/oleum from the various parts of the gas raising plant is collected in an intermediate tank with level control. The intermediate tank should be sufficiently large to contain about one week's production of acid. The collected acid is discharged intermittently (say once a week) with a membrane pump and is sent to a PTFE lined mixer, which is filled with glass balls. Here acid dilution with water takes place using an acid to water ratio of 1 : 10, so as to maintain the resultant sulphuric acid temperature below 50°C.

The resulting dilute sulphuric acid solution is transferred by gravity through Moplen (PVC) piping to a pit where the caustic solution of sodium sulphite (see 7.3.3.) from the SO_2 caustic scrubber is also discharged, so that neutralisation takes place. The pit contents (pH 8-10) can be used as a liquid component in slurry preparation for detergent powder manufacture.

Alternatively, the sulphuric acid/oleum may be piped at a low flow-rate directly to the reservoir of the caustic scrubber, providing that there is an interlock between the operation of the acid pump and the operation of the scrubber.

7.3.2 Treatment of ESP residues

The acid collected from an ESP (0.1 - 0.4% on organic feedstock) is in general dark coloured and may have a varying composition:

-	over-sulphonated organic acid	70-80%
-	sulphuric acid	20-30%

In the case of alpha-olefins, alcohols and alcohol ethoxylates high levels of decomposition products can also be present, notably high levels of 1,4-dioxane (0.5 - 1.0%) when alcohol ethers are sulphated and the residue remains in its acid form for long periods (hours). Residues from a plant running exclusively on alkylbenzene can be blended with the main production stream at a low level to meet sulphonic acid colour standards.

The ESP residues from all other organic feedstocks are not suitable for blending with normal production paste because of their high by-product levels and should be disposed of in a suitable way. ESP residues of feedstock other than alkylbenzene, notably residues from alcohol ether sulphation, should be neutralised immediately after collection to avoid the sharp increase in dioxane formation with time under extreme acidic conditions. The neutralisation should not take place in open vessels because of the risk of 1,4-dioxane liberation with subsequent potential operator exposure. There are various ways to solve the problem of disposal of the neutralised material:

(i) Specialist third party disposal: Waste producers are responsible to ensure waste is disposed of without harming operators, the public or the environment. This requires licenced disposal sites, competent and reliable waste contractor, proper labelling etc. For this solution no capital investment is required but the disadvantages are escalating costs, environmentalist pressure, declaration of waste composition and risk of accidents in transit.

(ii) On-site incinerating of acid residue: Certain specialist companies or companies for whom the acidic ESP residue represents a minor contribution to the effluent requiring incineration, prefer to incinerate the ESP residue on site.

7.3.3 Diluted caustic effluent disposal ex SO_2 scrubber

The liquid effluent leaving the SO_2 scrubber consists of a solution of caustic soda, sodium sulphite, sodium sulphate and traces of active matter. At 98% $SO_2 \rightarrow SO_3$ conversion (steady-state running conditions) about 80 kg/h of a 10% sodium sulphite solution will leave the scrubber system, based on a 1000 kg/hr sulphonation plant (see table 19). The alkaline solution is collected in an acid/caustic-resistant pit. It is common practice to dilute this liquid stream with other effluent streams (ex slurry-making and powder manufacture) and to re-use it in the detergent powder manufacturing plant as dilution water.

Due to the excess of caustic soda in the solution leaving the SO_2 scrubber (pH 9-12), this liquid can also be used to neutralise liquid acid effluents that cannot be disposed of in an SO_3 absorber and oleum/sulphuric acid ex the SO_3 cooler system. To avoid SO_2 formation due to decomposition of sodium sulphite (Na_2SO_3), the pH should never be lower than 8.

7.3.4 Summary table of various quantities of liquid effluents

Table 37 Liquid effluent quantities

Based on a sulphonation plant producing 1000 kg AM linear alkylbenzene sulphonate or 100 kg sulphur/hour (7% SO_3 in air).

1. Oleum from SO_3 cooler system

Assuming a dewpoint of -60°C, about 2 kg oleum/24 hours
Note: dewpoint air -40°C → 17 kg oleum/24 hours
-20°C → 138 kg oleum/24 hours

2. ESP residues

Organic acid/sulphuric acid
Quantities: 0.2 - 0.4% of organic feedstock depending on feedstock type and quality
Order of magnitude: 50 kg ESP residue/24 hours

3. Liquid effluent from SO_2 scrubber

At 98% SO_2 → SO_3 conversion (steady-state):

100/32 x 0.02 x 126 x 100/10 = 80 kg/hr or

at 50% SO_2 → SO_3 conversion (start-up):

100/32 x 0.5 x 126 x 100/10 = 2000 kg 10% Na_2SO_3/hr

7.4. Bibliography

1. Herman de Groot W. (1989). "Liquid effluents and exhaust gas emissions from sulphur based SO3/air sulphonation plants". Paper presented at Practical Sulphonation Seminar, 18-20 April 1989, The Hague, The Netherlands. Center for Professional Advancement, P.O. Box H, East Brunswick, New Jersey 08816, USA.

2. Ballestra SpA, Milano, Ballestra. Brochure 1990.

3. Chemithon Corporation, Seattle USA. "Hitec electrostatic precipitator".

4. Mazzoni SpA, Busto Arsizio, Italy. "SO_3 sulfonation/sulfation processing technology". Brochure (1988).

5. Meccaniche Moderne, Busto Arsizio, Italy.
"Sulfonation and sulfation plants". Catalogo A11/8/278.

8 PLANT START-UP/SHUT-DOWN AND OPERATIONAL CONDITIONS SULPHONATION/NEUTRALISATION

8.1 Plant start-up

Ensure availability at the various process units of the required utilities, set at the required process values and connected to the equipment items (example: cooling water flow, temperature and pressure to sulphonation reactor).

Ensure correct "routing" of the different raw materials from storage vessels and intermediate and final products to storage tanks (example: routing of the required type of organic feedstock to proportioning pump, from pump to reactor, from reactor discharge to start-up off-spec tank, etc., valve settings, level, temperature, pressure controls on).

Follow checklist of utilities and raw materials/products routing.

Switch on all panel instruments, alarms, etc.

Start all separate units which can be operated independently from the main process, i.e.:

* process air cooling system for air drying (freon compressor, glycol pumps, etc.);
* air-drier regeneration (stand-by);
* catalyst preheating;
* recycle pumps and cooling system for reactors (sulphonation, neutralisation) "ready for start-up";
* exhaust gas system: ESP on, H_2O/caustic feed to SO_2 scrubber, recycle pump on, etc.;
* start preheater burner, set relevant process temperatures;
* start process air blower, set relevant flows and pressures.

Ensure that the emergency system, dosing system, setting of flows, pressures and all other required functions/controls are <u>ready for start-up</u>.

When easy and logical procedures are defined there will be no need for special attention or parallel activities during the "critical" phase of production start-up. Therefore some actions are taken prematurely, in an order not exactly required at that time of the start-up procedures (example: caustic scrubber recycle and caustic feed, not necessary during plant preheating, will be started in advance to accumulate sufficient excess of caustic to absorb the initial extra amount of SO_2 produced during the initial start-up period after sulphur ignition).

Next start sulphur combustion and sulphonation reaction start-up.

Follow the step by step operating procedures. Then, once the required preheating temperatures are reached (normally within two hours), the plant will be ready for sulphur ignition: sulphur dosing and sulphur igniter will be started, the conversion of $SO_2 \rightarrow SO_3$ will proceed from an initial 50% value to 98% and the sulphonation reaction with alkylbenzene will result in a partially-converted alkylbenzene sulphonic acid, to be returned to the off-spec circulation tank. In this phase it is very important to follow a well-defined procedure and sequence of operation, in order to reduce the amount of off-spec material and start normal steady-state production leading to a product with the required specifications.

8.2 Plant shut-down

The philosophy of shut-down is to stop production in a way which minimises plant corrosion and product wastage. This is particularly true when the plant is to be shut down for more than 48 hours or when maintenance activities on the gas raising plant are envisaged. For short periods it is adequate simply to isolate the gas raising plant from the reactor. Therefore the following procedure is recommended.

A) Shut off the sulphur supply.

B) If any feedstock other than alkylate is currently being used, switch supply to alkylate and recirculate the alkylate continuously around the reactor until all the remaining SO_3 has been consumed.

C) Continue purging with dilution air to ensure the plant is free from residual SO_3 (30-60 minutes).

8.3 Operating conditions

8.3.1 Sulph(on)ation

The reaction of an organic feedstock with SO_3 in a falling-film reactor proceeds very quickly, with approximately 80% of the organic reacting in the first 1.5 to 2.0 metres of the 6 m-long reactor (see 5.5.3.). Associated with this quick reaction is a commensurate evolution of heat which must be dissipated equally as fast as it is produced to prevent hot spots leading to poor colour or even charring.

The film reactor design caters for this heat removal by utilising a split water cooling system to the reactor, allowing adjustment to water flow for the various types of feedstocks. LAB and lauryl alcohol require cooling water with typical temperatures between 20 and 30 °C. Cooling water loops with high recirculation rates are used. Fresh cooling water is fed to the loop to keep the recirculating cooling water at the pre-set value; 90% of the fresh cooling water is used in the top section of the reactor, with the remaining 10% used for the lower part of the reactor cooling circuit.

The combination of high cooling water recirculation and high rates of fresh cooling water results in a high heat-transfer coefficient and a high driving force for heat exchange, as the cooling water temperature rise between inlet and outlet in the top cooling circuit is approximately 1 °C. In the lower part, where far less heat is generated and where the viscosity of the organic becomes important, the cooling has to be tempered. This is accomplished by less water and allowing a rise of about 3 °C between inlet and outlet cooling water temperatures.

The SO_3 concentration is equally important to ensure product of acceptable colour and low free oil levels. The concentration of SO_3 in the gas phase is used to temper the reaction rate (see also 5.5.3.). For LAB the SO_3 concentration should be between 5 and 6% in air. Lauryl alcohol requires an SO_3 concentration of 4 to 5%, lauryl alcohol ethers 3%, and alpha-olefins 2.5 to 3%.

The final element necessary for efficient operation of the sulphonation reactor is the temperature of the feedstock. Generally the organic feed temperature should be as low as possible, for example between 25 and 28 °C for lauryl alcohol.

The sulphonation process conditions are summarised in table 38.

Table 38 Summary of sulphonation process conditions

Organic feedstock	SO_3 concentration % vol/vol	Organic feed temperature °C	Cooling water temperature °C
LAB	5-6	25-30	20-30
Lauryl alcohol	4-5	25-30	20-30
Ethoxylates	3	35-40	28-30
Tallow alcohol	4-5	55-60	45-50*
Alpha-olefins	2.5-3	10-15	12-16**
F.A. methyl esters	5-7	50-60	80-90

* Cooling water 2-3 °C above melting point of organic feedstocks to prevent freezing on cooling walls.

** Chilled cooling water required.

8.3.2 Neutralisation

It is essential to neutralise the acid forms of lauryl alcohol, alcohol ethoxylates and tallow alcohol as quickly as possible to prevent reversion (see also 4.3.3). To make "standard" concentrations of LABS, lauryl alcohol sulphate, lauryl alcohol ethoxylate sulphate and tallow alcohol sulphate, the following conditions are required (table 39):

Table 39 Neutralisation process conditions

Organic Feedstock	Cooling water °C	Organic acid from plant °C	Paste conc. % AM in paste
LAB	30 ± 5	40-48	45-60
Lauryl alcohol	30 ± 5	35-45	27-30, 70
Ethoxylate	30 ± 5	35-45	27-30, 70
Tallow alcohol	40-45	55	30, 60

8.4 Bibliography

1. Ballestra SpA, Milano. Technical proposal for sulphur based SO_3/air sulphonation/neutralisation plant with a multitube FFR sulphonation reactor and double step neutralisation reactor.

9 PLANT INSTRUMENTATION AND CONTROL*

The basic requirement of the sulphonation plant control system is to allow production of the desired quality acids and pastes in a safe manner within the design capacity limitations of the plant.
The level and sophistication of instruments and process control systems is largely determined by local preference although in order to satisfy the basic requirement, as stated above, there is a minimum requirement for instrumentation, control and safety interlocks which are common to all sulphonation plants.

* see also Process and Instrumentation Diagrams no. 1-8, at the end of this chapter.

9.1 Minimum instrumentation and control requirements

9.1.1 Air raising

Compressors

The outlet air temperature and pressure from the compressors gives a good indication of the state of downstream process plant, where additional pressure drop may be experienced due to the entrainment of dust in the drying bed, breakdown of the catalyst or malfunction of the control valves. The gas discharge pressure and preferably also the temperature should be recorded. In addition, for safety reasons and to protect the compressor, a high-pressure switch/alarm combination should be used to stop the compressor in the event of overpressure. The sulphur pumping system must be interlocked with compressor operation, to stop the pump when no process air is available.

Pressure control system

In order for the downstream control valves which control the flow-rate of SO_3/air to function in their optimum range and therefore enable the SO_3: organic mole ratio to be maintained as constant as possible, the air discharge pressure from the compressors must be controlled. Excess air from the compressors can be vented to atmosphere automatically through a valve controlled by the pressure measured at some point downstream of the venting point. An alternative system of controlling the compressor speed can be used where energy costs justify the additional capital expenditure.

Air coolers/chilling group

The temperature of the glycol solution used to cool the process air before entering the driers must be controlled. As an additional safeguard in the event that the temperature sensor or controller malfunctions, a low-temperature alarm, using a separate independent sensor, should be fitted on the glycol feed-line to warn if the temperature approaches $0^{\circ}C$. The consequence of too low a glycol feed temperature to the cooling air heat exchanger is the formation of ice on the finned pipes, leading to poor heat transfer, excessive pressure drop and possibly damage to the heat exchanger or its supports. A measurement of the total air flow-rate to the plant should be made on the process air directly after the chilling group. The mass-flow of air can be determined at this point by applying the appropriate temperature and pressure compensations to the volumetric flow-rate.

Air Driers

The control of the regeneration of drying beds can be a manual or automatic operation.

For automatic regeneration after fixed periods of operation, e.g. 8 hours, electromechanical (relays) or PLC based control may be used.

For plants without heat recovery to produce steam, hot air from sulphur combustion and SO_2/SO_3 cooling can be used for regeneration of the drying beds.

Dewpoint meter

A dewpoint meter for process air, with a recorded signal, should always be fitted as standard in sulphonation plants. Regular manual checks are essential (see 5.2.2.).

9.1.2 Sulphur handling

Proportioning

Molten sulphur must be delivered to the sulphur burner at a constant, desired rate to ensure that the subsequent downstream SO_3-in-air concentration and the SO_3 organic mole ratio are kept as constant as possible.

The commonest proportioning system for sulphur is to use calibrated variable-stroke metering pumps e.g. Bran & Luebbe. Normally the stroke length is adjusted manually by the operator and no system for closed-loop control of the sulphur flow-rate using metering pumps has ever been used in sulphonation plants. Closed-loop control of variable-speed gear pumps has however been used successfully, based on the measurement of the sulphur mass flowrate using flowmeters based on the Coriolis principle. The advantage of such a system is that the actual mass flow-rate is measured and controlled continuously.

Pressure measurement

Where pressure transducers or switches are used for molten sulphur, the sensor must be fitted with an extended diaphragm to prevent blockage resulting in false recordings.

Temperature control

The pressure of the steam used for heating sulphur must be rigorously controlled. High and low-pressure alarms should be installed in the feed lines, together with temperature indication.

9.1.3 Sulphur burning and SO_3 production

Ignition system

A sulphur ignition system is used in the Ballestra design of sulphur burner. The following safeguards must be incorporated:

- hard-wired interlocks between the sulphur igniter ammeter and operation of the sulphur pump;
- combustion air flow-rate and pressure indicators.

Sulphur burner

The temperature of the gases from the sulphur burner should always be recorded continuously and a high-temperature alarm fitted.

SO_2 coolers/catalyst tower/SO_3 coolers

Normally the temperature of the gases leaving the SO_2 coolers and entering the catalyst tower, the inlet temperature to each of the catalyst beds and the SO_3/air temperature leaving the SO_3 coolers are controlled by removing the heat of reaction using air-cooled heat exchangers or by the addition of cool process quench air. The control of the catalyst bed temperature profile by manipulation of the cooling air flow-rate may be performed manually or automatically. When the plant is expected to operate at several different sulphur flowrates due to capacity variation or due to the need to process several feedstocks of different molecular weights, the automatic temperature control systems are preferred to avoid the laborious task of resetting and rebalancing the cooling air flow-rates to each of the elements. For plants where operation is characterised by single feedstocks and long periods of steady-state output, manual systems are usually adequate. In either case, all, inter-stage temperatures of the SO_3 production plant should be recorded and a high-temperature alarm fitted after the final SO_3 cooler.

Oleum collection

The oleum collection vessel should be fitted with a means of determining the level (sight glass) and a high-level alarm to warn the operator of the need to empty the vessel and to give an indication, from records, of excessive oleum production. The discharge of oleum from the vessel is normally a manual operation, although in the case where oleum is pumped to the SO_2 scrubber, hard-wired interlocks must be used to prevent the transfer of oleum when the scrubber is not in operation.

9.1.4 Film Sulphonation

Mole ratio control

The effective control of mole ratio is the most important factor determining the quality of the sulphonated product.

The flow-rate of SO_3/air is controlled by modulating valves linked to a flow sensor, normally an orifice plate. Depending on the required SO_3-in-air concentration for the sulphonation of different feedstocks, it is necessary to dilute the SO_3 from the gas-raising plant with dried process air. The central system for dilution air again employs a flow sensor and modulating valve.

Accurate metering of organic material to the reactor is of critical importance. As with control of the sulphur flow-rate, proportioning pumps or gear pump/mass flowmeter combinations can be used.

At present, Ballestra has installed automatic control of the SO_3/organic mole ratio in at least six new sulphonation plants. The principles of the mole ratio control are indicated in P&I diagram 8. This new development is feasible because accurate and technical reliable massflow meters are available now.

Reactor cooling system

The temperature of the cooling water to the reactor must be controlled. It is normal to pump cooler water to the upper sections of the reactor, where most of the heat is produced, which therefore requires separate cooling water temperature control. The temperature of the cooling water entering and leaving the reactors should be measured and, if necessary, the flow-rates adjusted such that the ΔT does not exceed 1-2°C. The flow-rate of cooling water to individual sections of the reactor and the product outlet temperature should also be measured. The consequence of ineffective cooling is poorer quality products.

Reactor emergency systems

Hard-wired interlocks should be provided to prevent opening of the SO_3/air isolation valve until the reactor emergency system is primed. The system should be activated by a hard-wired interlock to an organic flow sensor.

Level control - acid/gas separator

Control of the level in the acid/gas separator ensures an effective seal between the liquid handling parts of the sulphonation plant and the exhaust gas treatment section. Several methods can be used, but the one which minimizes the hold-up and residence time distribution of acids, and consequently dioxane formation during ether sulphate manufacture is based on control of a variable speed gear pump.

Ageing vessel/hydrolyser

It is essential to cool the ageing vessel and hydrolyser during the manufacture of alkylbenzene sulphonic acids. Although manual control of the cooling water flow-rate to the ageing vessel in particular is often adequate, a flow indicator should be installed in the cooling water line and a temperature indicator in the product outlet line to enable the operator to adjust the cooling water flow-rate effectively. Where strict acid colour specification must be adhered to, the installation of an automatic temperature control system may be justified.

9.1.5 Neutralisation

Metering systems

Sodium hydroxide or other neutralising agents, water and other solvents if required are normally added to the neutralisation loop using metering pumps. Variable-speed gear pumps controlled by mass-flow meter-based systems can be used.

pH control

Closed loop control of pH is required. Where metering pumps are used for reagent addition, automatic stroke adjustment should be used to control the addition of a small fraction of the neutralising agent, the bulk of it being added by a manually adjustable metering pump.

Pressure control

The operating pressure of the neutralisation loop is normally not controlled. However, a high pressure switch must be incorporated to stop the paste recirculation pump in the event of overpressure.

Temperature control

In neutralisation the temperature of the product is not controlled directly, as the feed of "too cold" water to the shell of the heat exchanger can result in the formation of a skin of semi-solid product on the inner surfaces of the tubes. The cooling water temperature is controlled instead, by the addition of cold make-up water to the cooling water recirculation system. This method of control can only be used if the heat exchanger is oversized for the maximum envisaged duty with reference to a minimum cooling water temperature in the recirculation cooling water loop of 25°C. In tropical locations, where the make-up water may be at a higher temperature than the minimum as defined for "skin" formation on the heat exchange tubes, the heat exchanger must be sized for the actual characteristics of the cooling water supply.

9.1.6 Exhaust Gas Treatment

SO_2 scrubber

In order to ensure that the pH in the SO_2 scrubber does not fall excessively, leading to gas emission, it is essential to control the addition of sodium hydroxide to the unit. Cascade control systems can be used, where the pH transducer is used to define a set point for a flow control loop. However, variable-speed metering pumps are more commonly used, the speed of the pump being controlled directly by pH.
A second requirement of the scrubbing control system is to maintain the total solids level in the effluent below 10% w/w. The single variable-speed metering pump motor can therefore be used to drive a second head for the addition of water, thereby controlling the correct water : sodium hydroxide ratio. The temperature and flow-rate of the recirculation liquor should be measured and a low-flow alarm incorporated in the liquor recirculation line.

It is normally required to install an SO_2 monitor in the exhaust gas stack.

9.2 Process control strategy

The level of sophistication of the process control strategy is largely determined by local preference. Several options are available.

9.2.1 Manual control

In the case of manual control, the operator follows written start-up, shut-down and product changeover procedures. During normal steady-state operation key variables (section 9.1.) are controlled automatically, usually by panel-mounted PID controllers. The operator makes adjustments to set points of the controllers and also adjusts pump stroke lengths in response to the results from plant-side analysis or in response to abnormal readings from local or panel-mounted instruments. This type of control strategy is best suited to plants where long periods of steady-state operation on single products are envisaged, with few, if any, product changeovers.

9.2.2 Assisted manual control/data logging

Manual control of the operation of a sulphonation plant can be supplemented by computer-based data logging. Such systems are non-interactive in that analogue and digital signals from the plant are fed to the computer/data logging system, but information is not fed back. Consequently, the status of pumps etc. and other data can be displayed on the screen-based mimics, reports can be printed and the data obtained can be manipulated to show trends or used in other off-line computer programs.

9.2.3 Supervisory system

Unlike simple data logging systems, supervisory control systems are designed to restrict the flexibility of the operator by "supervision" of his actions. In these systems the operator can interact with the plant through the keyboard/workstation to start or stop pumps, open valves and change PID controller set points. The three-term controllers may be conventional electronic controllers, linked to the supervising system, or software algorithms programmed into a PLC.

The sophistication of the supervisory control system depends somewhat on the level of plant automation. In principle, all valves on the plant could be remotely actuated, allowing almost all procedures including start-up and shut-down to be "run" by the computer. In practice, such levels of automation are not cost-effective and a compromise is normally reached on the level of plant automation.

The level of automation depends largely on site preference and a high level of automation is normally only justifiable when several products at different capacities are manufactured on a single plant. In this case it is important to ensure correct routing of feedstock to reactor and product to storage. Routing of this type can be incorporated into logic checks within the supervisory system. Even in this case, valve automation may not be justified and manual valves fitted with proximity switches can be used to inform the system that the operator has actuated the correct valves.

If automatic valves are used, however, they can be actuated automatically from sequences in the system initiated by the selection of a particular "recipe" by the operator. Such recipes may contain not only routing information but also all the information required to produce a particular product, starting from a particular feedstock. This information could be transferred to the plant either by the operator or by the system, dependent upon the level of automation.

The main advantage of sophisticated supervisory control systems is that, by restricting the activities of the operator, mistakes during production which would affect safety or product quality can similarly be restricted. However, by restricting the operator's actions, a measure of flexibility is lost in the manner in which the plant can be operated. This loss of flexibility is often important when elements in the plant are not functioning as intended. For example, if it is not possible to use the dedicated storage tank for a particular product which must be produced and an equivalent storage tank is available, then re-routing the product to the new tank not only requires valve position changes but also software changes before the system will accept the new configuration. Although most systems can be re-configured in a straightforward manner, such action is normally outside the scope of the operator's capabilities and intervention from higher-level supervisors or technicians is required.

On the basis of the foregoing statements, the usual control philosophy adopted when designing a supervisory control system is as follows:

a. computer-assisted start-up, shut-down and product changeover;
b. fully automated continuous running;
c. automated emergency shut-down (usually hard-wired).

a. Computer assisted start-up

The basic principle for a computer-aided start-up (shut-down) is to present the operator with a checklist of procedures that he or she must undertake either just prior to or during the 2 hour pre-heat cycle of the gas-raising plant. Where possible the computer will cross-check the operator actions by:

- monitoring routing valve positions from proximity switches;
- measurements of temperature, pressure, flow, etc.
- ensuring logical sequences are followed, e.g., pre-heat catalyst bed before starting the sulphur pump.

It is estimated that these checks could take the operator 2 hours. The principal advantages of this approach are:

- limited expenditure on automation, i.e. routing valves etc. not used during continuous production operations;
- restrict the complexity.

b. Automated continuous running

All the valves, motors, pumps and analogue loops in the direct process stream are controlled via a computer based system. Services and utilities are only monitored and alarmed. Quality measurements either in-line or off-line are monitored by the system. As all the necessary information is available to the computer, it can calculate and perform:

- material balances/consumption;
- conversion factors;
- adjustments to optimise processing;
- alarm on drift from pre-set operating conditions.

From start-up (sulphur ignition) the plant is fully computer controlled and operated for a given (pre-set) type of product and production rate. The change of product type and/or production rate is still performed by the operator and computer-assisted in terms of control and relevant process parameters.

c. Automatic emergency stop

The approach is not different from the traditional control system where fail safe and plant alarm/interlocks can be brought into action at a moment's notice so as not to endanger life or plant.

9.3 Bibliography

1. Ballestra SpA, Milano. Technical proposal for sulphur based SO_3/air sulphonation/neutralisation plant with a multitube sulphonator reactor and double step neutralisation reactor.

2. Davidson J.G. Private communications, 1989 and 1990.

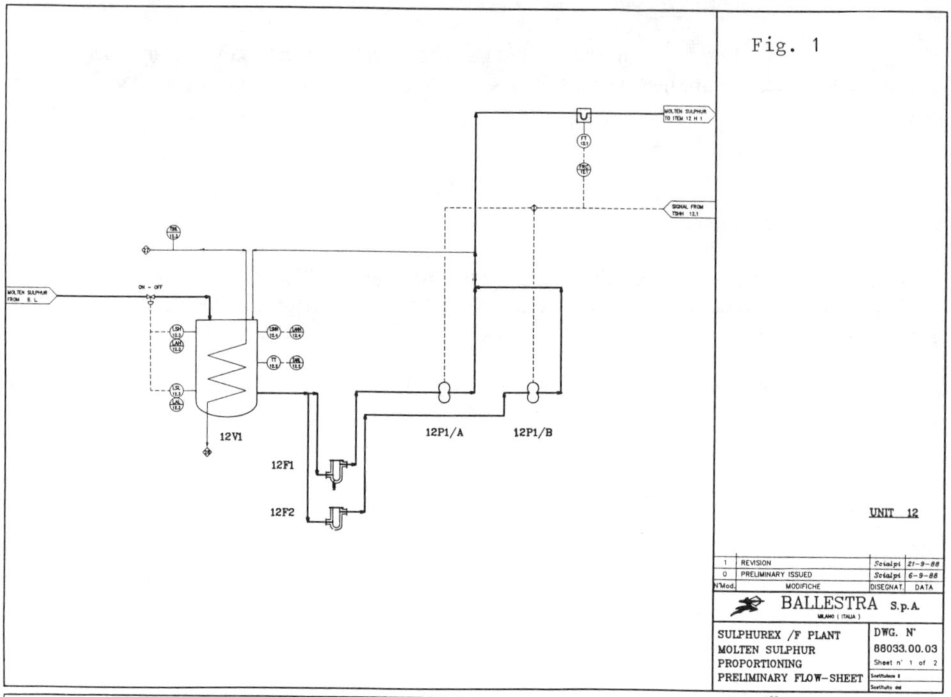
MOLTEN SULPHUR TO ITEM 12 H 1
SIGNAL FROM TSHH 12.1
MOLTEN SULPHUR FROM B. L.
ON - OFF
12V1
12P1/A
12P1/B
12F1
12F2
UNIT 12
1 REVISION Scialpi 21-9-88
0 PRELIMINARY ISSUED Scialpi 6-9-88
N°Mod. MODIFICHE DISEGNAT. DATA
BALLESTRA S.p.A.
MILANO (ITALIA)
SULPHUREX /F PLANT
MOLTEN SULPHUR
PROPORTIONING
PRELIMINARY FLOW-SHEET
DWG. N°
88033.00.03
Sheet n° 1 of 2

Fig. 1

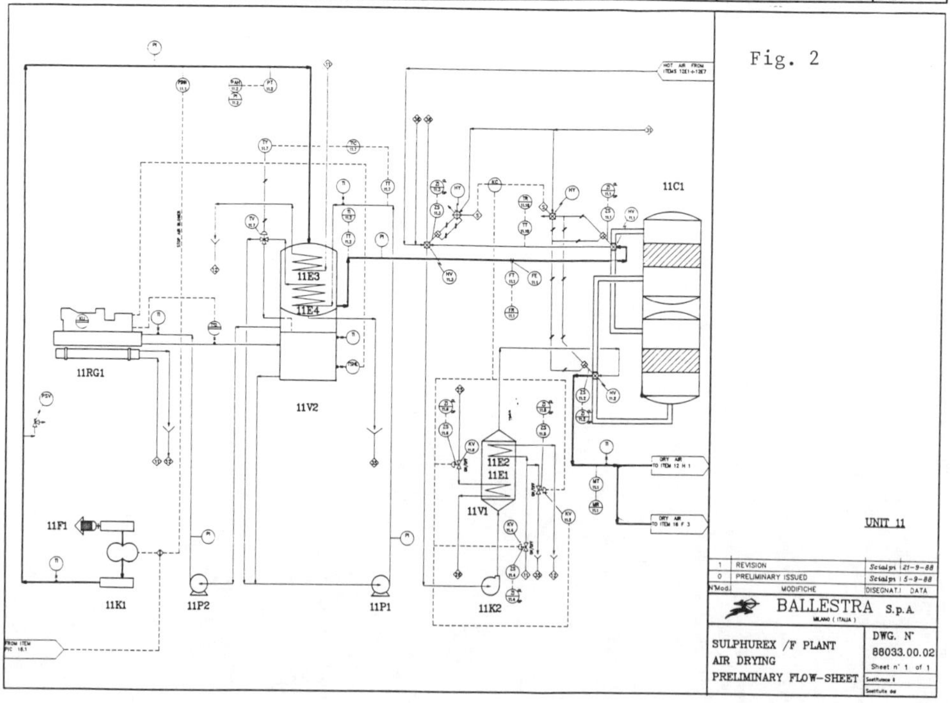
HOT AIR FROM ITEMS 12E1÷12E7
11C1
11E3
11E4
11RG1
11V2
11E2
11E1
11V1
11F1
11K1
11P2
11P1
11K2
DRY AIR TO ITEM 12 H 1
DRY AIR TO ITEM 18 F 3
FROM ITEM PIC 18.1
UNIT 11
1 REVISION Scialpi 21-9-88
0 PRELIMINARY ISSUED Scialpi 5-9-88
N°Mod. MODIFICHE DISEGNAT. DATA
BALLESTRA S.p.A.
MILANO (ITALIA)
SULPHUREX /F PLANT
AIR DRYING
PRELIMINARY FLOW-SHEET
DWG. N°
88033.00.02
Sheet n° 1 of 1

Fig. 2

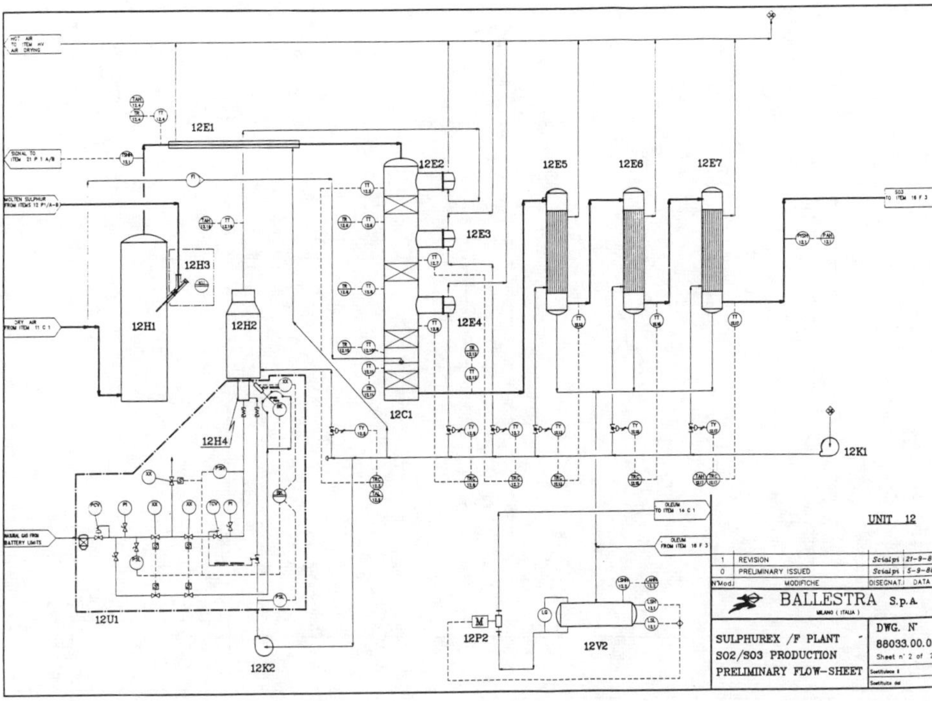
12E1
12E2
12E3
12E4
12E5
12E6
12E7
12H1
12H2
12H3
12H4
12C1
12U1
12K1
12K2
12P2
12V2
UNIT 12
1 REVISION Sciaipi 21-9-88
0 PRELIMINARY ISSUED Sciaipi 5-9-88
N°Mod. MODIFICHE DISEGNAT. DATA
BALLESTRA S.p.A.
SULPHUREX /F PLANT
SO2/SO3 PRODUCTION
PRELIMINARY FLOW-SHEET
DWG. N°
88033.00.03
Sheet n° 2 of 2

Fig. 3

Fig. 4

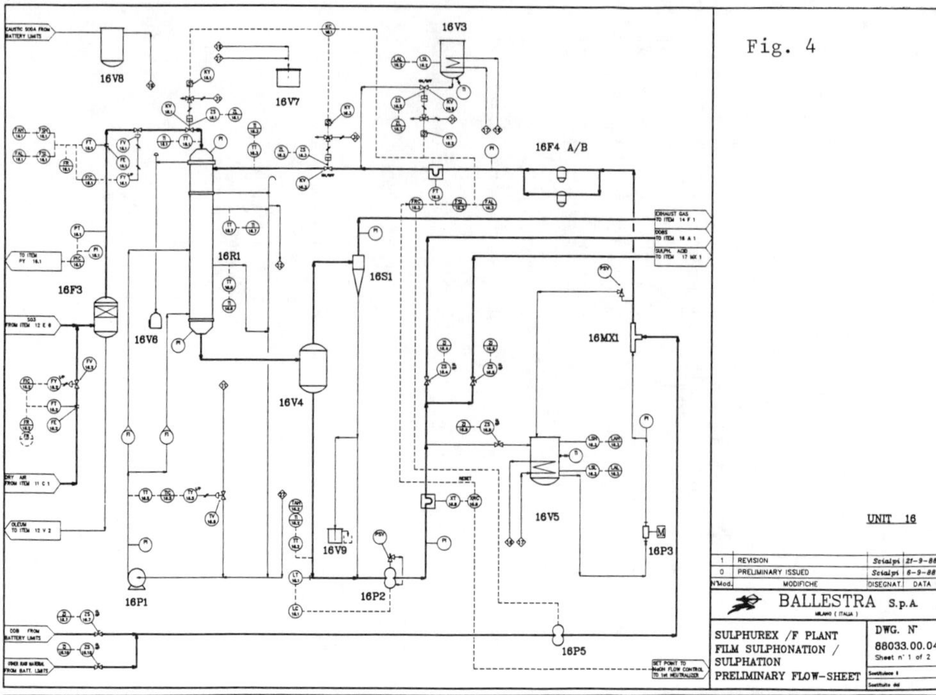

Fig. 4
(continued)

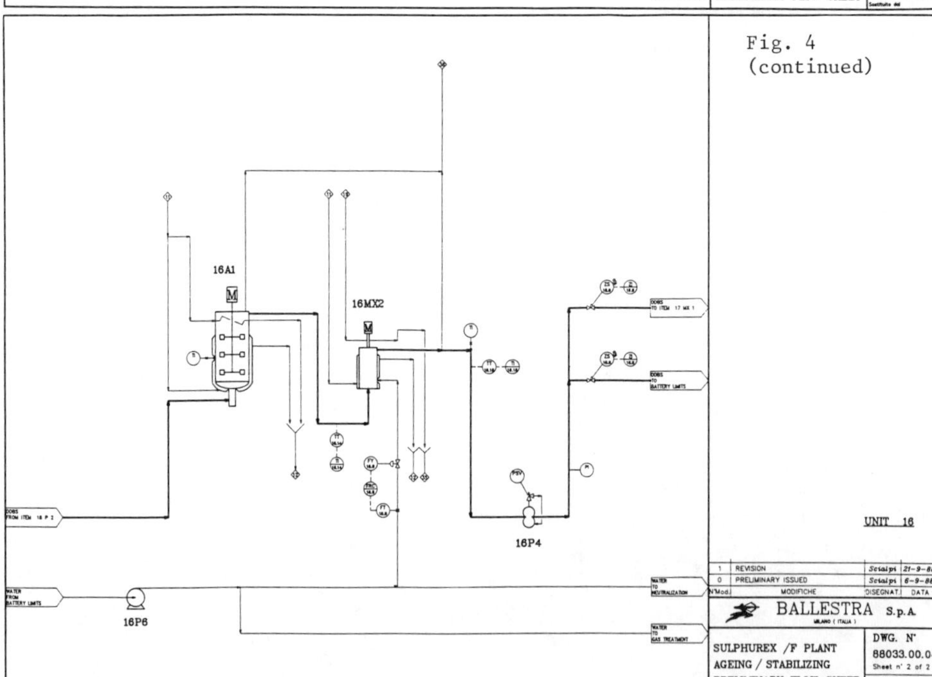

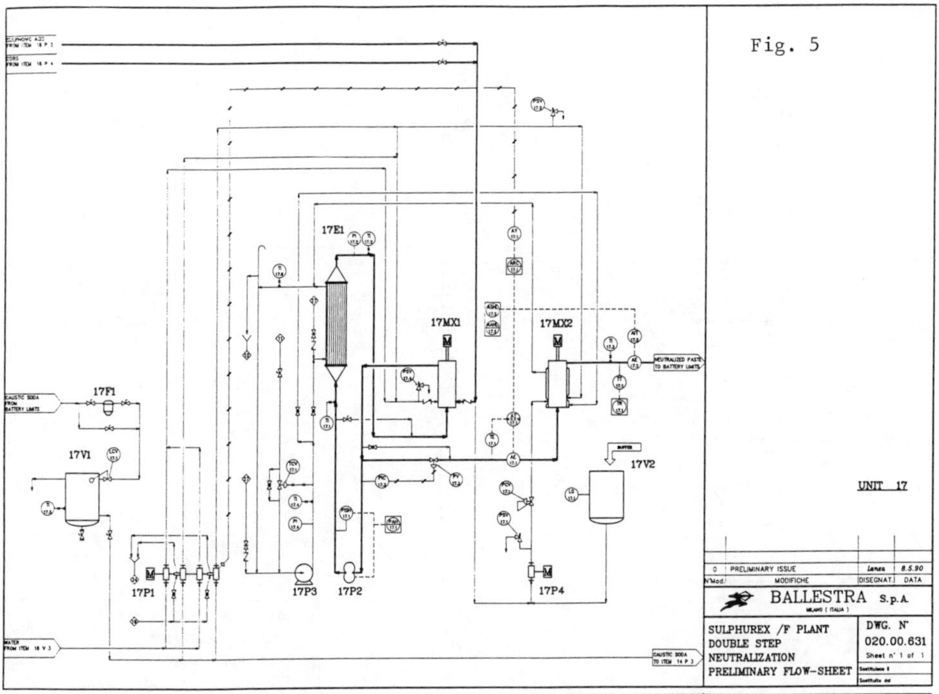
17E1
17MX1
17MX2
17F1
17V1
17V2
17P1
17P2
17P3
17P4
UNIT 17
0 PRELIMINARY ISSUE
BALLESTRA S.p.A.
SULPHUREX /F PLANT
DOUBLE STEP
NEUTRALIZATION
PRELIMINARY FLOW-SHEET
DWG. N°
020.00.631

Fig. 5

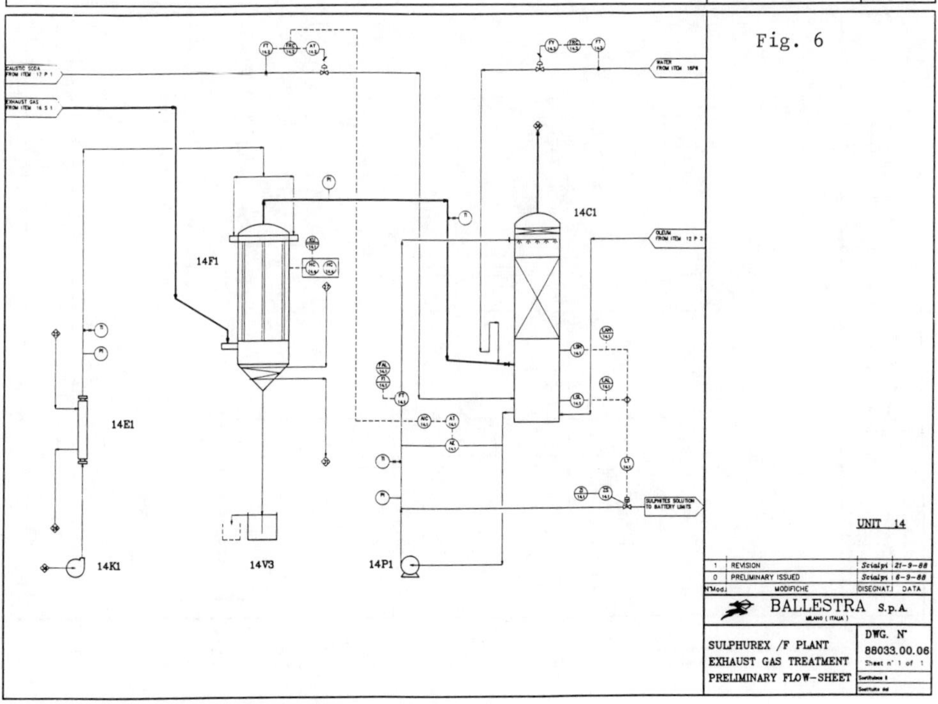
14C1
14F1
14E1
14K1
14V3
14P1
UNIT 14
1 REVISION
0 PRELIMINARY ISSUED
BALLESTRA S.p.A.
SULPHUREX /F PLANT
EXHAUST GAS TREATMENT
PRELIMINARY FLOW-SHEET
DWG. N°
88033.00.06

Fig. 6

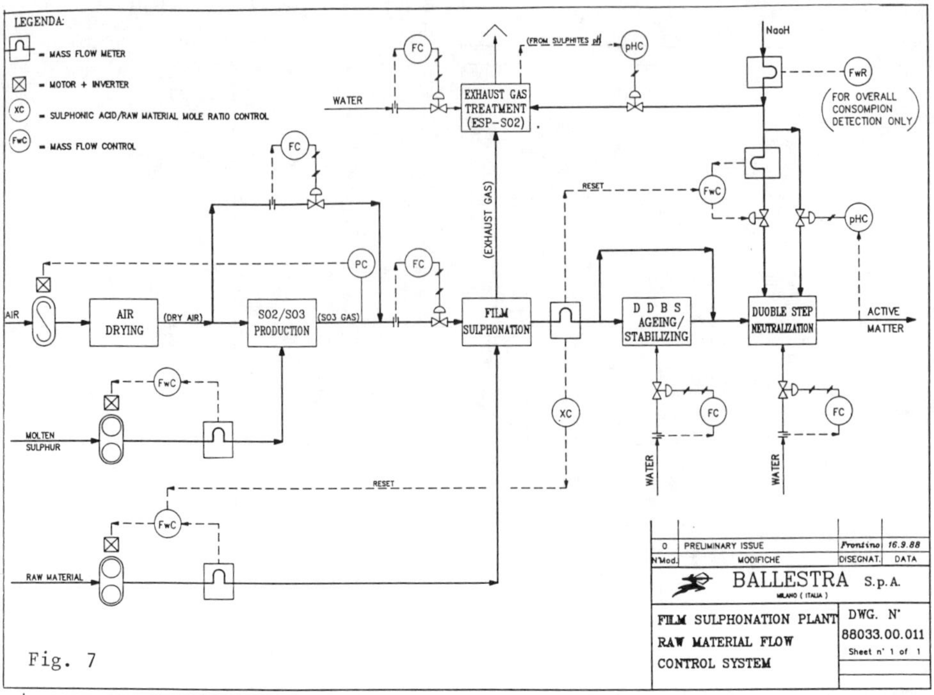
LEGENDA:
= MASS FLOW METER
= MOTOR + INVERTER
XC = SULPHONIC ACID/RAW MATERIAL MOLE RATIO CONTROL
FwC = MASS FLOW CONTROL
WATER
EXHAUST GAS TREATMENT (ESP-SO2)
(FROM SULPHITES pH)
NaoH
FOR OVERALL CONSOMPION DETECTION ONLY
RESET
(EXHAUST GAS)
AIR
AIR DRYING
(DRY AIR)
SO2/SO3 PRODUCTION
(SO3 GAS)
FILM SULPHONATION
D D B S AGEING/ STABILIZING
DUOBLE STEP NEUTRALIZATION
ACTIVE MATTER
MOLTEN SULPHUR
RAW MATERIAL
0 PRELIMINARY ISSUE Frontino 16.9.88
N°Mod. MODIFICHE DISEGNAT. DATA
BALLESTRA S.p.A.
MILANO (ITALIA)
FILM SULPHONATION PLANT RAW MATERIAL FLOW CONTROL SYSTEM
DWG. N° 88033.00.011
Sheet n° 1 of 1

Fig. 7

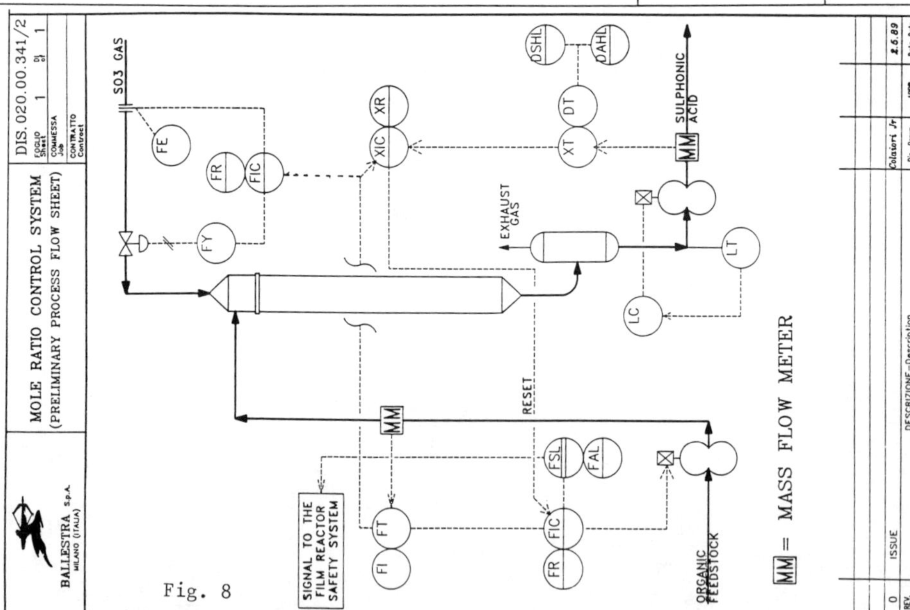
MOLE RATIO CONTROL SYSTEM (PRELIMINARY PROCESS FLOW SHEET)
DIS. 020.00.341/2
FOGLIO Sheet 1 DI of 1
COMMESSA Job
CONTRATTO Contract
BALLESTRA S.p.A. MILANO (ITALIA)
SO3 GAS
EXHAUST GAS
SULPHONIC ACID
RESET
SIGNAL TO THE FILM REACTOR SAFETY SYSTEM
ORGANIC FEEDSTOCK
MM = MASS FLOW METER
Colaiori Jr
2.6.89
0 ISSUE
REV. DESCRIZIONE-Description Dis.-Drawn APPR. Data-Date

Fig. 8

10 PLANT LOCATION, LAYOUT, BUILDING STRUCTURE, PLANT DOCUMENTATION

10.1 General considerations

Owing to the hazardous nature of the materials handled in a sulphonation plant it should be accommodated in its own building at least 20m from other buildings. It should not be in the centre of the factory but on the periphery, remote from residential and other areas where gas emissions could be considered unsafe or a nuisance. The site should have easy access by road for the receipt of raw materials and should be near the powders/liquids processing plants so that long transfer lines for sulph(on)ated products (acids or pastes) can be avoided.

Sulphur powder can be stored in air open bunkers provided that they have a water-tight roof to keep out the rain. Molten sulphur storage tanks and organic feedstock, sulphonic acid and AD paste tanks can also be located in the open air provided that the local climate is not extremely cold.

The whole of the gas raising plant can be sited in the open air, dependent upon the climate and the plant throughput. Smaller plants (1-2 ton AD/h throughput) normally have indoor gas-raising plants, notably in tropical areas and in very cold regions of the globe. The effluent treatment plant can be outside the building.

10.2 Plant layout

The equipment suppliers quote the following overall dimensions for typical plant layouts (table 40).

Table 40 Overall dimensions for typical plant layouts

Plant cap.	1.5 ton AD/h				3.0 ton AD/h			
	Length (m)	Width (m)	Height (m)	Floor area (m^2)	Length (m)	Width (m)	Height (m)	Floor area (m^2)
Ballestra	24	18	11	432	36	18	13.5	648
MM	34	22	10	748	40	27	10.5	1080
Chemithon	33	17	9	561	36	23	10	828

The plant dimensions in table 40 are indicative. Examples of plant layout are depicted in figures 49, 50 and 51.

Fig. 49. TYPICAL SULPHONATION PLANT LAYOUT

BALLESTRA SULPHUREX PLANT 2000 kg/h

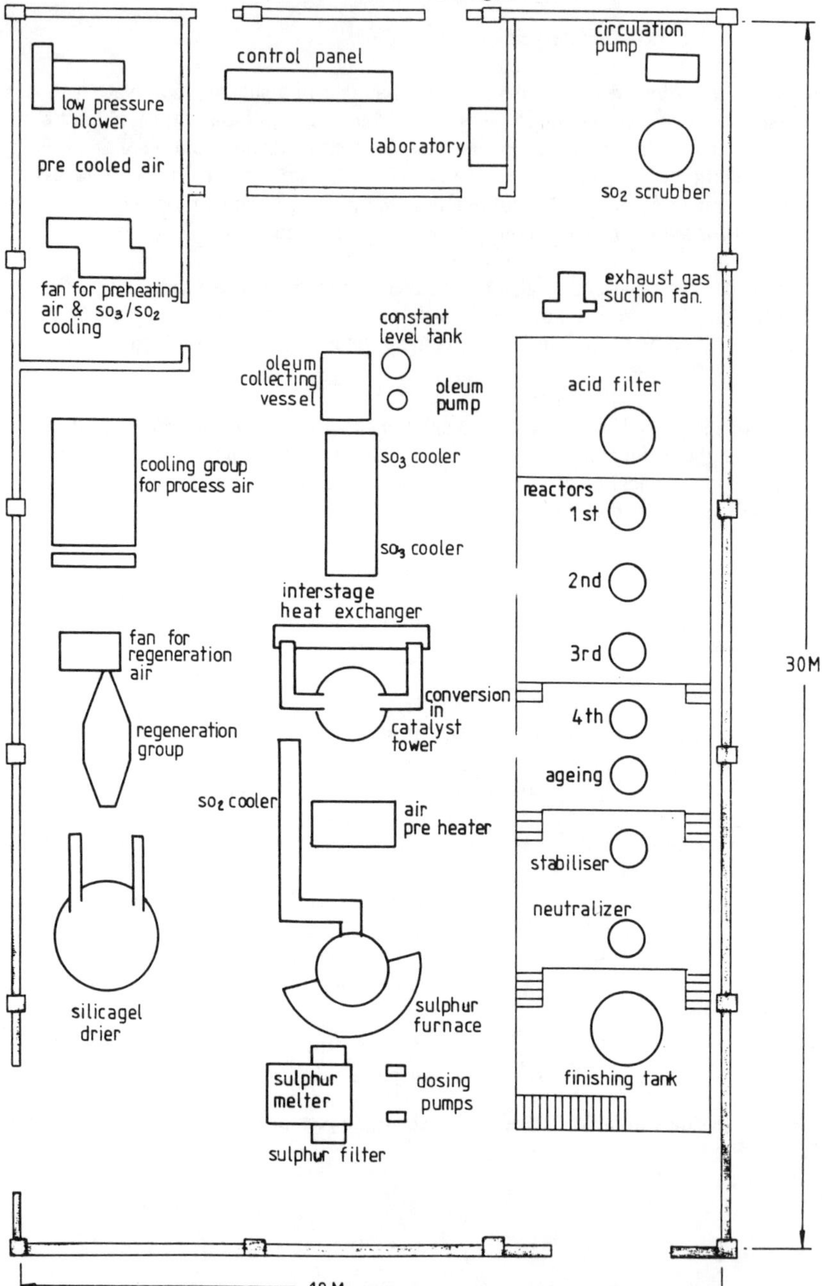

Fig. 50. TYPICAL SULPHONATION PLANT LAYOUT
MECCANICHE MODERNE 3000 kg/h

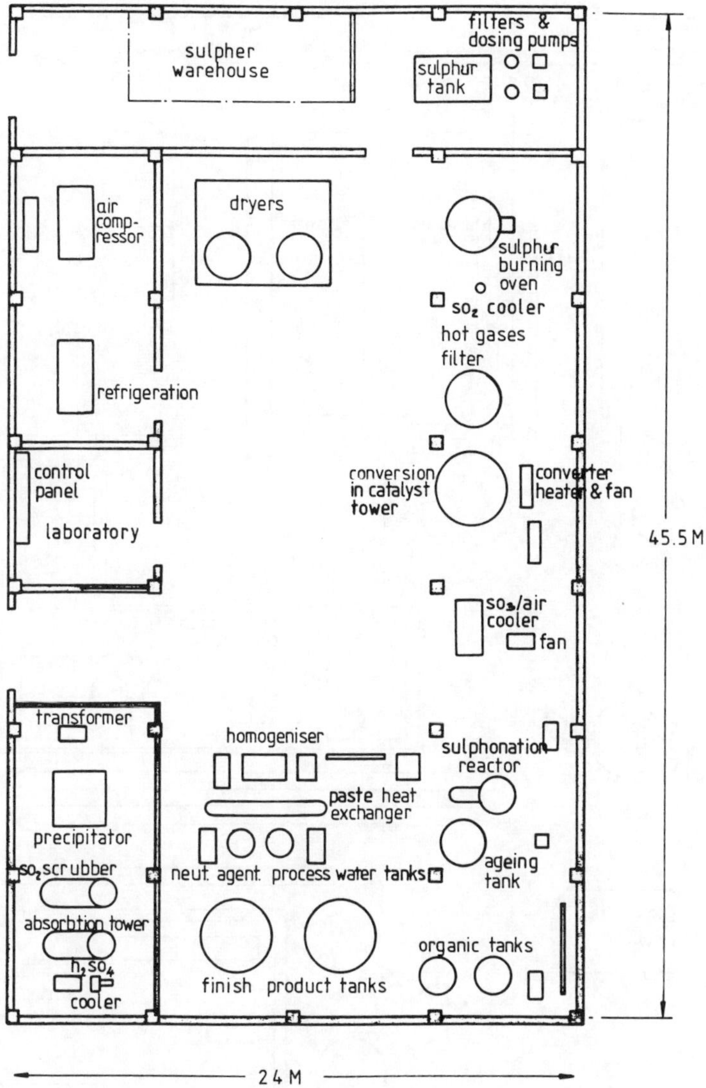

Fig. 51. TYPICAL SULPHONATION PLANT LAYOUT
CHEMITHON 2000 kg/h

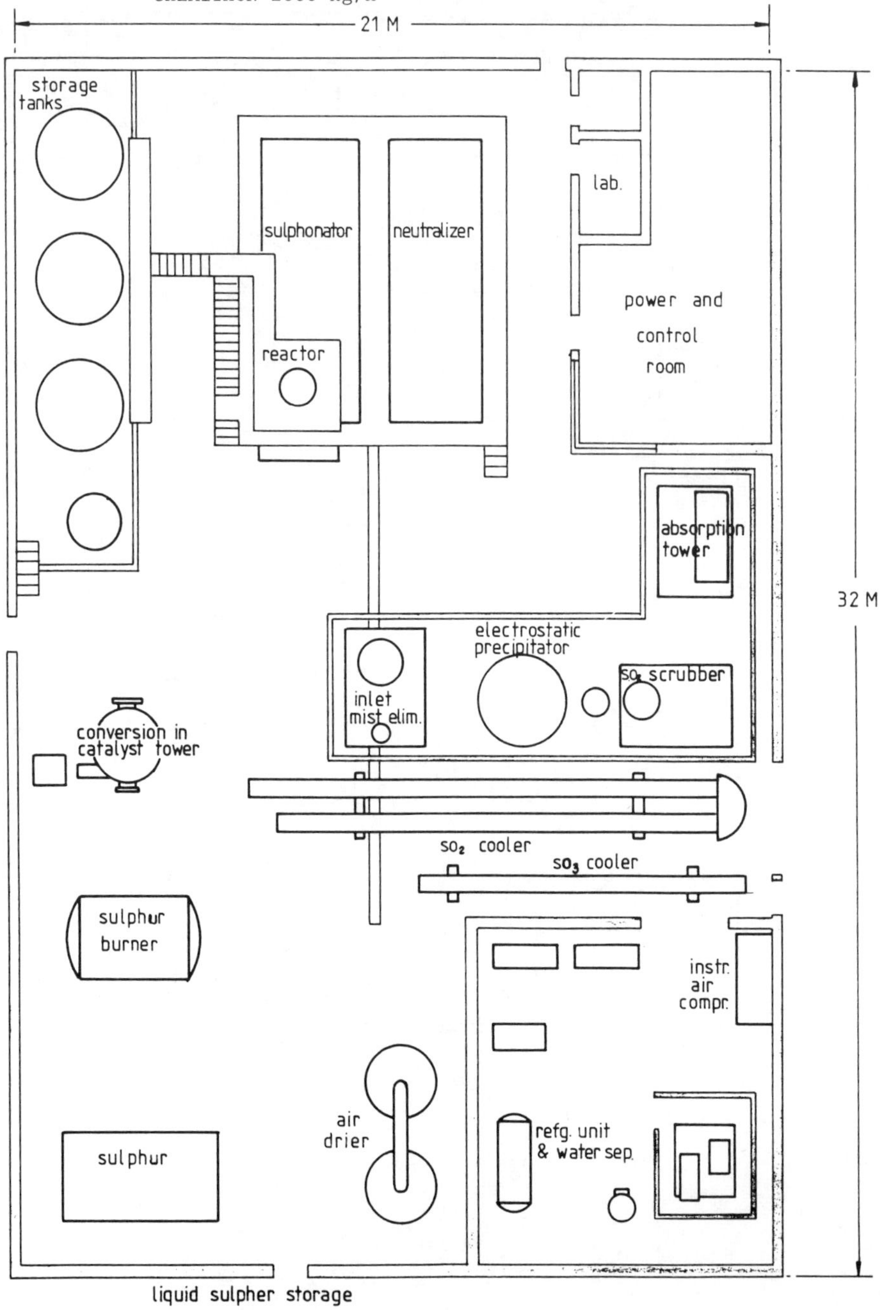

The following general pattern in plant layout can be observed:

- an air-drying "line";
- an SO_3/air gas-raising "line";
- a sulphonation, neutralisation and exhaust air treatment "line";
- the control room, a separate room with control mimic panel or computer system with VDU and printer and motor control centre and separate ventilation system from the plant areas. The control room should have access from both plant areas (convenience) and outside the building to enable operators to enter the room in the event of a gas release in the processing area;
- a small laboratory adjacent to the control room, where product samples can be analysed. It is therefore recommended to position the control room/laboratory in the proximity of the sulphonation/ neutralisation units;
- all noisy equipment should be sited in a separate brick-built room to contain:

(i) the process air compressor(s). Consideration should be given to enclosing the process air compressor(s) in its own separate acoustic enclosure, the objective being to reduce the noise level outside the compressor/blower room to less than 78 decibels, one metre from the brick wall;
(ii) the cooling air blower;
(iii) the refrigeration unit for process air drying.

10.3 Building structure

The design of the building depends on the climatic conditions. A suitable structure might be a portal frame filled with block or brick side walls to a height of 3 m with aluminium sheeting above. A lifting beam is required over the sulphonation reactor, vessels and ESP. It is convenient to arrange access through a large window in the roof for both installation and major maintenance of heavy equipment. The plant area should be large enough for ample access by fork-lift truck for maintenance and repair work. The floor under the SO_3/air coolers, ESP and scrubber must be covered with heavy-duty (35 mm) acid resistant tiles cemented with "Furane" cement or equivalent.

Equipment must be raised off the floor on plinths and acid-proof drains should discharge into the pit outside the building where neutralisation of the plant effluents can take place prior to discharge to the factory drain or process water tanks.

The building should include a separate brick-built room for the sulphur melter and filters when solid sulphur is used.

Exposed steelwork used in the construction of the building should be continuously joined by welding or by jumper cables welded to bridge individual components. The whole should be earthed to a suitable ground mat or ground rods. Day tanks for organic feedstocks or products, equipment frames and the ESP must be similarly earthed.

Safety showers and eye baths should be installed in suitable locations around the plant. These items are sometimes installed after plant commissioning is completed, whereas, due to the nature of commissioning, they may be most required <u>during</u> the commissioning phase. It is therefore essential to consider the supply and installation of safety showers and eye-baths as an intrinsic part of the plant.

10.4 Plant documentation

Proper plant maintenance is of great importance for plant lifespan and product quality. Equipment suppliers should deliver detailed written instructions for maintenance of plant items and plant maintenance procedures.
An operating instructional manual should be prepared by the equipment supplier before the actual plant becomes operational. This book should contain the following sections:

1. Introduction
   - scope of project
   - ambient conditions
   - utilities consumptions
2. Local Feedstock(s) specifications
3. Chemicals specifications e.g.
   - sulphur
   - caustic
   - ammonia
   - alcohol
4. Product quality guarantees
5. Process description
   - general descriptions related to P&I flow sheets for various sections
   - lay-out
   - typical desired plant operation conditions
   - mass balances related to local feed-stock(s)
   - inventory of plant, list of components and valves
6. Plant operation
   - preliminary operations
   - start-up procedures
   - shut-down procedures
7. Emergencies
   - powercuts
   - no cooling water
   - no caustic
8. Trouble shooting
   - (see appendix 5)
9. Analytical checks
   - analytical tests required in plant
   - method of sampling
   - where to sample
   - test frequency

		- range of acceptability
10.	Plant safety	- list of hazardous chemicals - physiological effects - first-aid treatment

There should also be prepared a mechanical/instrumental catalogue, containing specific drawings and/or data about plant items like compressors, chiller unit, dosing pumps, transfer pumps, main equipment items like sulphur burner, converter tower, drying equipment, reactors, mixers, etc. etc., instruments and control system.

Proper plant documentation is essential in supervisor/ operator training, longterm plant operation and good maintenance of the total plant.

11 SULPHONATION WITH 20% OLEUM (see also 2.3)

11.1 Chemistry of sulphonation with 20% oleum, mass balances and heat balances

The organic feedstock, alkylbenzene, is reacted with a 10-15% excess (by weight) of 20% oleum to ensure high conversions in a short time. As the reaction is both rapid and highly exothermic, efficient heat removal is required to prevent oversulphonation and product darkening.

The main reaction can be depicted as follows:

$$RC_6H_5 + H_2SO_4.SO_3 \rightarrow RC_6H_4SO_3H + H_2SO_4 + H_2O$$
$$\Delta H = -112 \text{ kJ/gmole}$$

Secondary reactions may take place:

$$RC_6H_5 + H_2SO_4.SO_3 \rightarrow \underset{\text{disulphonic acid}}{RC_6H_4\,(SO_3H)_2} + H_2SO_4 + H_2O$$

$$RC_6H_4SO_3H + RC_6H_5 \rightarrow \underset{\text{sulphone}}{(RC_6H_4)_2SO_2} + H_2O$$

The undiluted mixture from the sulphonation stage typically contains 60 - 70% alkylbenzene sulphonic acid, 30 - 40% sulphuric acid, about 1% water and up to 1% free oil. For the purpose of determining what separation will take place, the small amount of free oil may be neglected. Without free oil, a product post-sulphonation loop could have, for example, the following composition:

62.7% ABSA, 36.2% H_2SO_4 and 1.1% H_2O

Figure 52 presents a phase diagram for the system alkylbenzene sulphonic acid - sulphuric acid - water. Before dilution, the sulphonation mixture is represented by a point (P) close to the bottom of the diagram, near or on the lower perimeter of the area of immiscibility. In diluting with water the composition of the sulphonation mixture "moves up" into the area of immiscibility: see dotted line P-L. After dilution into the area of immiscibility, the mixture of alkylbenzene sulphonic acid, sulphuric acid and water will separate into two phases: a spent acid phase with very little sulphonic acid (about 0.3%) and a sulphonic acid phase. The composition of the latter is that of the intersection of the appropriate tie line (say the tie line indicated as 80% spent acid composition) with the left hand perimeter of the area of immiscibility (point R). The composition of the spent acid phase is given by the intersection of the same tie-line with the zero percent sulphonic acid line (see point S), since the small amount of sulphonic acid in the spent acid phase is neglected for the purpose of this phase diagram. From S, Q and R, all on the 80% spent acid tie-line, the composition of the sulphonic acid phase and the amounts of sulphonic and spent acid phases can be calculated.

Different alkylbenzene qualities may present different phase diagrams and it may be necessary to determine the phase diagram in laboratory trials to arrive at the optimal conditions, i.e. a sulphonic acid phase with the minimum amount of sulphuric acid and water.

Fig. 52. Phase-diagram (60°C)
Sulphonic acid - Sulphuric acid - Water

A Area of immiscibility

B Homogeneous - phase area

C Area of jelly formation

D Area of aqueous solution

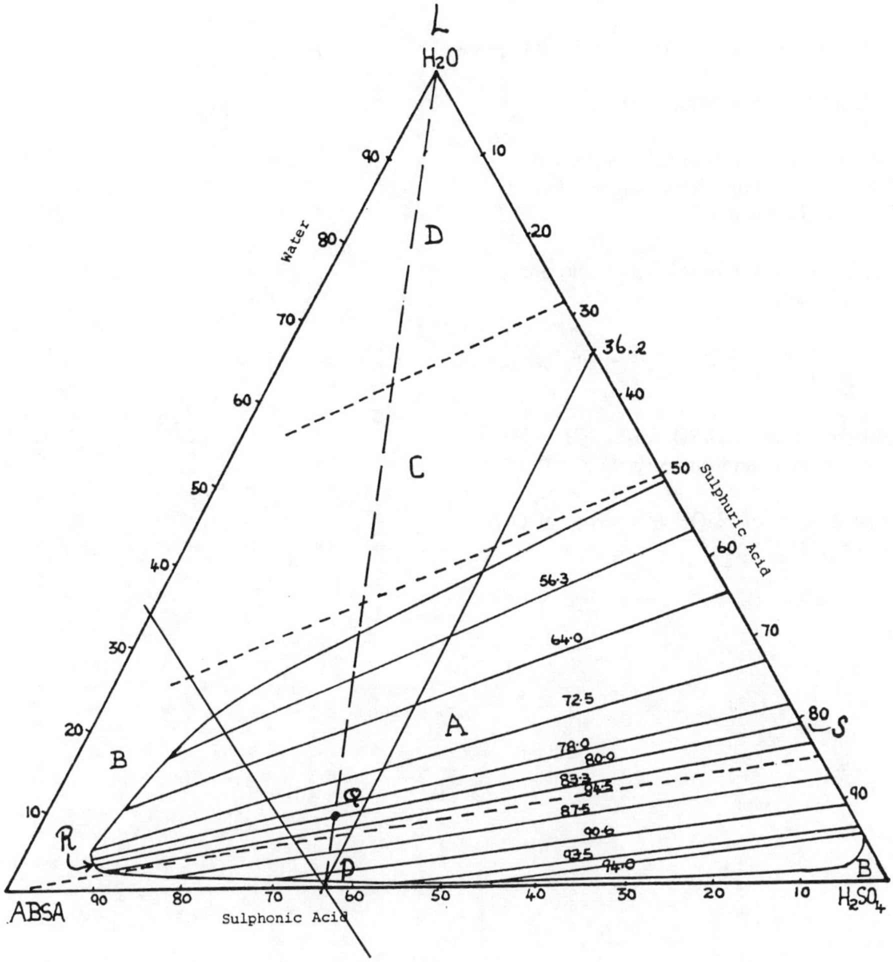

In this section a typical mass balance is given for the 20% oleum sulphonation reaction. The total result for the mass balance is shown in table 41. All the data shown are accurate to about 0.5 kg and are based on the following typical values of the various parameters concerned:

oleum strength (by weight)	20% SO_3
oleum/alkylbenzene ratio by weight	1.1/1.0
sulphonic acid yield	98.5%
% free oil on a 100% AD basis in final paste	1.1
molecular weight of alkylbenzene	244
{100% x (H_2SO_4 from sulphonation stage)}/ {(H_2O from sulphonation stage + H_2O as fresh dilution water)}	76
H_2SO_4 carry-over in sulphonic acid phase (weight %)	8.5
sulphonic acid carry-over in spent acid (weight%)	0.3
concentrations of H_2SO_4 in (H_2SO_4 + H_2O) part of spent acid phase (weight %)	78
concentration of NaOH in aqueous NaOH solution (weight %)	14.5

Table 41 Typical 20% oleum sulphonation plant. Mass balance based on the assumptions listed in section 11.1. All units in kg.

	%		1000 AD	1500 AD	2000 AD	2500 AD	3000 AD
Raw Materials		alkylbenzene	717.5	1076.0	1435.0	1793.5	2152.0
		20% oleum	789.0	1183.5	1578.0	1973.0	2367.0
		dil. H_2O	154.0	231.5	308.5	385.5	463.0
		NaOH	191.0	1540.0	381.5	477.0	572.5
		neutr. H_2O	1125.0	434.0	2250.5	2813.0	3375.5
		Total	2976.5	4465.0	5953.5	7442.0	8930.0
From Sulphonation Stage	62.3	ABSA	938.5	1407.5	1876.5	2346.0	2815.0
	0.7	NDOM*	10.5	16.0	21.5	27.0	32.0
	35.9	H_2SO_4	541.0	811.5	1082.0	1352.0	1622.5
	1.1	H_2O	16.5	24.5	33.0	41.0	49.5
		Total	1506.5	2259.5	3013.0	3766.0	4519.0
From Dilution Stage		ABSA	938.5	1407.5	1876.5	2346.0	2815.0
		NDOM*	10.5	16.0	21.5	27.0	32.0
		H_2SO_4	541.0	811.5	1082.0	1352.0	1622.5
		H_2O	170.5	256.0	341.5	427.0	512.5
		Total	1660.5	2491.0	3321.5	4152.0	4982.0
From Separation Stage (i) sulphonic acid phase	86.4	ABSA	936.5	1405.0	1873.0	2341.5	2809.5
	1.0	NDOM*	10.5	16.0	21.5	27.0	32.0
	8.5	H_2SO_4	92.0	138.5	184.5	230.5	276.5
	4.1	H_2O	44.0	66.0	88.0	110.5	132.5
		Total	1083.0	1625.5	2167.0	2709.5	3250.5
(ii) spent acid phase	0.3	ABSA	2.0	2.5	3.5	4.5	5.5
	77.8	H_2SO_4	448.5	673.0	897.5	1122.0	1346.5
	21.9	H_2O	126.5	190.0	253.0	316.5	379.5
		Total	577.0	865.5	1154.0	1443.0	1731.5
		(recy. ABSA)	(0.5)	(1.0)	(1.5)	(2.0)	(2.0)
		(recy. H_2SO_4)	(177.5)	(266.0)	(354.5)	(443.5)	(532.0)
		(recy. H_2O)	(50.0)	(75.0)	(100.0)	(125.0)	(150.0)
		(Total)	(805.0)	(1207.5)	(1610.0)	(2013.5)	(2415.5)
From Neutralisation Stage	41.7	Na-ABS	1000.0	1500.0	2000.0	2500.0	3000.0
	0.4	NDOM*	10.5	16.0	21.5	27.0	32.0
	52.3	H_2O	1255.5	1883.0	2511.0	3138.5	3766.0
	5.6	Na_2SO_4	133.5	200.5	267	333.5	400.5
		Total	2399.5	3599.5	4799.5	5999.0	7198.5

* NDOM = free oil

All reactions (sulphonation, dilution and neutralisation) are highly exothermic, with the following ΔH values:

sulphonation	ΔH = - 112 kJ/gmole AB
dilution	ΔH = - 79 kJ/gmole AB
neutralisation	ΔH = - 502 kJ/kg acid mixture

The heat of neutralisation of 502 kJ/kg acid mixture is based on an acid composition of 86.4% ABSA, 1.0% free oil, 8.5% H_2SO_4, 4.1% H_2O, when the sulphonation is carried out at 50°C with a 20% oleum/AB ratio of 1.10 (weight %).

Table 42 Typical values that occur for various heat transfer parameters in 20% oleum sulphonation, dilution and caustic soda neutralisation

	1000 kg AD/h	2000 kg AD/h	3000 kg AD/h
Sulphonation			
Temperature of sulphonation product (°C)	50	50	50
Overall heat transfer coeff. (W/m² °C)	350	350	350
Heat transfer area heat exchanger (m²)	10.7	21.3	32
Temp. driving force (°C)	20	20	20
Heat transferred (kW)	74.7	149	224
Temp. cooling water in cooling loop (°C)	30	30	30
Temp. fresh cooling water to loop (°C)	20	20	20
Required amount of cooling water (kg/s)	1.78	3.55	5.33
Dilution			
Temp. of dilution product (°C)	63	63	63
Overall heat transfer coeff. (W/m² °C)	120	120	120
Heat transfer area heat exchanger (m²)	15	25	35
Temp. driving force (°C)	26.5	32	34
Heat transferred (kW)	46	92	138
Temp. cooling water in cooling loop (°C)	36.5	31	29
Temp. fresh cooling water to loop (°C)	20	20	20
Required amount of cooling water (kg/s)	0.66	2.0	3.65
Neutralisation (with NaOH)			
Temp. of neutralisation product (°C)	64	64	64
Overall heat transfer coeff. (W/m² °C)	300	300	300
Heat transfer area heat exchanger (m²)	10.7	21.3	32
Temp. driving force (°C)	27	27	27
Heat transferred (kW)	86.7	173	259
Temp. cooling water in cooling loop (°C)	37	37	37
Temp. fresh cooling water to loop (°C)	20	20	20
Required amount of cooling water (kg/s)	1.21	2.42	3.63

From heat balance calculations it is possible to determine the heat load (i.e. heat transferred in the heat exchangers of the sulphonation, dilution and neutralisation loops) at each stage of the total process, and hence the amount of cooling water needed. The results of these calculations are summarised in table 42. The recirculation ratio (volume of recirculating material divided by the volume of throughput) is 15:1 to give small temperature differences in the sulphonation, dilution and neutralisation loop process-side streams.
Cooling water is also circulated at a high rate in a cooling water loop, fed with fresh cooling water from a cooling tower circuit. Therefore the temperature driving force is roughly equal to the difference between process loop and cooling water loop temperatures.

11.2 20% oleum plant equipment (see figure 53)

Sulphonation

An oleum sulphonation system consists of three stages: sulphonation, dilution/separation and neutralisation. The flows of alkylbenzene and oleum are automatically controlled into the suction side of the centrifugal circulation pump of the sulphonation loop reactor. The liquids are injected through an injection tube consisting of two concentric tubes made of polypropylene and incoloy Alloy 825, almost to the entrance of the impeller to ensure that mixing takes place in the pump and not earlier. The circulation rate of the circulating liquid is about 15 times the sum of the rates at which alkylbenzene and oleum are introduced. This ensures that the heat released during the reaction is absorbed by a large quantity of liquid, and thus there is only a small rise in temperature. Also, very thorough and rapid mixing is ensured. The shell-and-tube sulphonation heat exchanger removes the heat of reaction and the heat generated by the centrifugal circulation pump. The reaction mixture leaves the heat exchanger at a constant temperature of about 50°C. From the sulphonator, the sulphonation product passes to a post-reactor hold-up coil with a mean residence time of 5 minutes.

The arrangement of the loop reactor (with perfect mixer characteristics) and the post-reaction coil (approaching plug flow characteristics) will make sure that the maximum conversion of the LAB feedstock is achieved. All equipment is constructed in stainless steel 316L.

Dilution

The dilution loop consists of a centrifugal circulation pump with impervious graphite housing and impervious graphite heat exchangers. All pipework should preferably be of mild steel lined with polytetrafluorethylene (PTFE) because of the corrosivity of the acid mixture, containing 80% sulphuric acid.

The impervious graphite heat exchangers (e.g. Vicarb or le Carbone Lorraine) consist of several graphite blocks with two cooling water headers of steel on each block. The graphite headers and blocks are held together between two end-plates which are connected by rods. Two sets of channels have been drilled perpendicularly to each other through each block, the acid mixture passing through one set and cooling water through the other. Heat is transferred through the graphite (high heat conductivity) between the channels. An automatically controlled flow of dilution water is injected almost to the entrance of the impeller to ensure that mixing with the circulating flow takes place in the pump and not earlier. The circulation rate of the circulating liquid is, like in the sulphonation loop, about 15 times the sum of the rates at which the acid mixture and water are introduced.

Separation

The separator is a vertical cylindrical vessel made from Werkstoff 4505 (comparable to STST316L but containing 2% Cu, 1.5% N_b and higher in N_i than 316L) or constructed of polyester resin reinforced with glass fibre. This is protected on the inside by a 3 mm layer of PVC, chemically bonded to the glass fibre. Although the latter separators are cheaper than the Werkstoff type, special skills would be needed to repair them if this became necessary.

A sight glass is built into the front of the separator and should extend from about one quarter of the height from the bottom to about one quarter of the height from the top. When the plant is in operation, the lower part of the separator is filled with spent acid. Above this is an intermediate layer, in which separation occurs and above the intermediate layer is sulphonic acid. The dark spent acid layer is distinct from the rest of the separator contents, as is visible through the sight glass.

The acid mixture from the diluter is introduced through a distributor made of polypropylene, located in a horizontal plane in the middle of the separator. The volume of the sulphonic acid layer is equal to that of the spent acid layer. The distributor is constructed so that turbulence in the liquid is avoided.

The sulphonic acid phase leaves the separator at the top, the spent acid phase at the bottom. Good separation between the sulphonic acid and the spent acid is important, in order to minimise the loss of sulphonic acid in the spent acid phase. Also, the higher the sulphuric acid content of the sulphonic acid phase, the less acceptable the latter is for many of its applications. Furthermore, the higher the sulphonic acid content of the spent acid, the less acceptable this is to its prospective buyers, because of foam formation in further processing.

The cross-sectional area of the separator should be 1 m^2 per ton AD/h. To ensure good separation, the sulphonic acid phase should remain in the separator for about 60 minutes. All the pipework should preferably be made of mild steel lined with PTFE.

Neutralisation

The neutralisation stage consists of a centrifugal circulation pump (e.g. KSB type CPK-C) and a heat exchanger made of stainless steel 316L. A sodium hydroxide solution (e.g. 14.5% NaOH) and the sulphonic acid from the separator enter the neutralisation loop reactor via the circulation pump. Entry is through an injection tube consisting of two concentric tubes made of polypropylene. The recycle flow rate in the loop is the same as that in the sulphonation and dilution stages. The heat evolved during neutralisation is removed in the neutralisation heat exchanger. A neutralised paste is obtained containing about 40% sodium alkylbenzene sulphonate. The paste from the neutralisation loop passes on to the paste receiving tank.

Automatic pH control of the paste from the neutralisation stage works as follows:

A sample of paste is automatically taken, diluted with water by a factor of ten and the pH continuously measured. The measured pH value influences the setting of the pH control valve, which is connected in parallel to the main flow control valve for sodium hydroxide. The control loop for the main flow control valve has to be set to the amount of NaOH necessary for neutralisation, and fine dosing takes place via the pH control valve in the pH control loop.

Raw material metering

Centrifugal pumps with flowmeters can be used for the dosing of oleum, alkylbenzene, dilution water and caustic solution. Positive displacement pumps are more accurate, but also more expensive. Furthermore, when a positive displacement pump is used, there must be a pressure relief system to prevent the pressure in the discharge line becoming too high and there must also be a pressure damper in the discharge line.

Acid storage (20% oleum and spent acid)

Oleum should be stored in thick-walled mild-steel tanks. As there is considerable risk of corrosion at the gas-liquid interface in an oleum tank, care should be taken to admit only dry air to the tank when it is being emptied. During filling, SO_3 vapours should preferably be absorbed from displaced air. Therefore, oleum tanks are vented to the atmosphere via a sulphuric acid scrubber.
The spent acid from the separator is stored in tanks having a lining of either stainless steel or polyvinyl chloride (PVC).
There should always be an empty stand-by emergency tank for 20% oleum and spent acid.

Tandem sulphonation procedures

Chemithon are known to have "tandem" sulphonated mixtures of alkylbenzene/primary alcohols. The process is similar to the normal operation. In the first loop 20% oleum and alkylbenzene are dosed, in the second loop (the diluter) the alcohol with extra oleum is introduced, followed by immediate neutralisation (see figure 54). A product mix of ABS/alcohol sulphate/sodium sulphate is finally obtained.

11.3 **Product quality**

The quality of alkylbenzene sulphonate from 20% oleum plants is normally very good. The free oil figures fall in a range between 0.6 and 1.7% on 100% AD. Due to the dissolved sulphuric acid in the sulphonic acid phase, the sodium sulphate levels on 100% AD vary between 6 and 15% depending on the type of alkylate and the efficiency of the separation process of sulphonic acid/spent acid. Paste colour is normally very good, with values well below 100 Klett (5% AD solution, 4 cm cell, F42 filter).

Figure 53 Chemithon 20% oleum sulphonation plant

CHEMITHON

SCHEMATIC FLOW DIAGRAM

CONTINUOUS DETERGENT SLURRY PROCESSING PLANT

HIGH-ACTIVE ALKYLATE SULFONATION WITH OLEUM

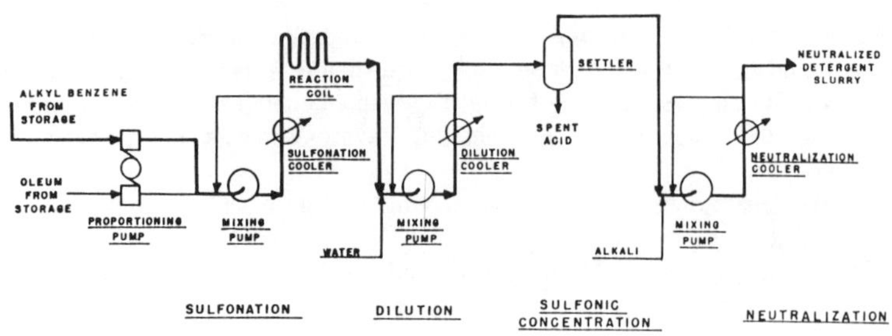

Figure 54 Chemithon 20% oleum tandem sulphonation plant (LAB and PA)

CHEMITHON

SCHEMATIC FLOW DIAGRAM

CONTINUOUS DETERGENT SLURRY PROCESSING PLANT

TANDEM TYPE SULFONATION-SULFATION WITH OLEUM

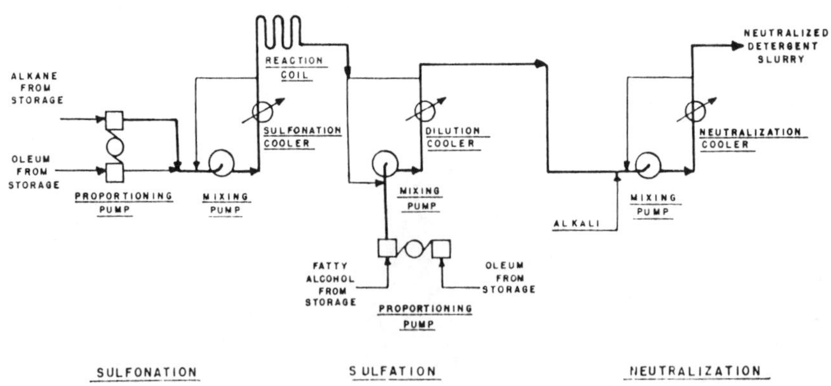

11.4 Bibliography

1. Chemithon Corporation, Seattle, USA.
"The Chemithon* process and equipment for continuous oleum processing of detergent raw materials", appr. 1960, 26 pages.

2. Chemithon Corporation, Seattle, USA.
5430 West Marginal Way, SW, Seattle, WA 98106 USA.
Chemithon sulfonation/sulfation brochure, 1988.

3. Falbe J. (editor) (1986). Surfactants in consumer products. Theory, technology and applications. Springer Verlag, Berlin.
Section 3.1.1 "General production technologies", pp 54-62.

4. Le Carbone Lorraine, 45 Rue des Acacias, 75821 Paris. Brochures about Graphilor*, (impermeable graphite), block heat exchangers and centrifugal pumps.

5. Le Carbone Lorraine, Brochures about ARMYLOR G*(PTFE). ARMYLOR G* is used to protect the internal surfaces of metal piping (carbon steel) against attack by chemically corrosive process fluids.

12 SULPHONATION TECHNOLOGY AS A TOOL FOR THE EDUCATOR IN PREPARING CASE STUDIES AND FINAL-YEAR DESIGN PROJECTS IN A CHEMICAL ENGINEERING CURRICULUM

12.1 Introduction

This section outlines the virtues of the sulphonation technology as a "learning paradise" for chemical engineers-to-be. All kinds of problems and disciplines come together in one concise operation: highly exothermic reactions, sometimes in combination with substantial viscosity increase of the reaction products, undesired parallel and consecutive reactions affecting final product quality, environmental problems, hazardous and corrosive chemicals demanding careful selection of plant equipment, design and construction materials.

This section is meant to help the teacher to prepare case studies to illustrate general chemical engineering theory and/or to prepare a more extensive briefing for a final-year Design Project. The author of this book could act as an outside consultant in the event that some extra help or checks on results are required. (Einhoven University of Technology, Chemical Engineering Department (T), Laboratory of Chemical Process Technology (TCP) - P.O. Box 513 - 5600 MB Eindhoven, The Netherlands - telephone (31) 40 47 29 70 (secretary), telefax (31) 40 44 25 76).

12.2 Mass balance exercise (sulphur-based SO_3/air sulphonation, 100 kg S/h)

The following data are supplied:

1. **Process air drying**

Inlet air (20 °C)	2130 kg/h
H_2O in inlet air	19.3 kg/h
Air temperature after chiller (see 5.2.2.)	3 °C
Air dewpoint after silica gel drier (see 5.2.2.)	-60 °C

2. **Sulphur burner** (see 4.1.1.)

Sulphur (100%) to furnace	100 kg/h
Air to burner	A kg/h
Air from burner	B kg/h
SO_2 from burner (100% combustion of S)	C kg/h

3. **From SO_2 → SO_3 converter** (see 4.1.2.)

Total conversion SO_2 → SO_3	98%
Air from converter	D kg/h
SO_3	E kg/h
SO_2	F kg/h

4. **Oleum (20%) condensation in SO_3 coolers** (see 5.2.7.)

20% oleum collected	negligible

5. **To sulphonator**

Organic feedstock LAB (MW 237)	710 kg/h
Air	D kg/h
SO_3	E kg/h
SO_2	F kg/h

6. **From ageing and stabiliser** (see 4.2.1. and 5.5.10.)

Sulphonic acid (100%) (note 9, sulphonic acid antrainment)	G kg/h
free oil (see 4.2.1.)	10.0 kg/h
H_2SO_4	5.1 kg/h
Water	8.6 kg/h

7. **To neutraliser** (see 4.3.2.)

NaOH (50% W/W strength)	H kg/h
Dilution water	I kg/h
Sulphonic acid mix	(G+10.0+5.1+8.6) kg/h

8. **From neutraliser**

100% active matter (AM)	J kg/h
paste strength % (100% AM weight/ total weight)	55%
free oil	10.0 kg/h
Na_2SO_4	K kg/h
Total H_2O in paste phase	L kg/h

9. **To ESP** (see 5.9.1., 7.2.2., 7.2.3.)

Air	D kg/h
SO_2	F kg/h
Sulphuric acid mist (20% oleum)	2.9 kg/h
Sulphonic acid entrainment	3 kg/h

Assume that sulphuric acid mist and organic mist are completely trapped in ESP.

10. **To SO_2 scrubber**

Air	D kg/h
SO_2	F kg/h

Assume complete absorption of SO_2 in scrubber.

From stack

Air	D kg/h

Problems to be solved

---

(1) Calculate the amount of water removed in the process air chilling unit.
(2) Calculate the amount of water removed in the silica gel bed drier (air with a dew point of -60°C contains 0.009 gram water per kg dry air).
(3) Calculate the following mass flows: A, B, C, D, E, F, G, H, I, J, K, L.
(4) Prepare a sulphur mass balance.

12.3 Mass and heat balance exercises - equipment design tasks (all based on Massbalance exercise 12.2)

12.3.1 Sulphur furnace (see 4.1.1.)

Calculate the outlet temperature of the SO_2/air mixture leaving the sulphur furnace.

12.3.2 SO_2/air cooler (see 5.2.4.)

Design a double hairpin ("trombone") cooler to cool the SO_2/air ex sulphur furnace to a temperature of 430°C with ambient air flowing counter-current through the jacket of the trombone cooler; inlet cooling air temperature 30°C, outlet 250°C. Mean SO_2/air flow velocity in tube 25 m/s, mean cooling air velocity in jacket 20 m/s.

Problems to be solved

---

(1) Prepare a heat balance for the SO_2/air - cooling air heat exchanger. Use Cp (kJ/kg°C) values of mean temperatures process side, cooling air side. From the heat balance, the mass flow of the cooling air will be found.
(2) Calculate tube and jacket diameters.
(3) Calculate heat transfer coefficients, hi and hu and take into account the radiation contribution in hu. From hi and hu the overall heat transfer coefficient, U (W/m^2°C), can be calculated.
(4) Calculate the length of the trombone cooler.
(5) Calculate the lagging of the double hairpin heat exchanger assuming a heat conductivity coefficient, k, of the glass wool insulation material of 0.15 W/m°C and a heat transfer coefficient ha (W/m^2°C) from the Al-cladding to the ambient air of 25 W/m^2°C.
(6) Make a sketch of the double hairpin cooler with expansion pieces.
(7) What construction materials would you select for the SO_2/air tube and the cooling air jacket?

12.3.3 $SO_2 \rightarrow SO_3$ converter (see 4.1.2. and table 4)

Assume 74% conversion in upper bed of catalyst tower.

(1) Calculate the temperature of the gas mixture leaving the first bed. Use mass balance data from 12.2. and use mean Cp values in relevant temperature range.
(2) Design a shell-and-tube heat exchanger (SO_2/SO_3/air through tubes, cooling air in shell, cooling air in 30°C, cooling air out 250°C) between the upper and second catalyst beds. Inlet temperature SO_2/SO_3/air second bed 440°C. Use relevant literature for design of shell-and-tube heat exchangers. Take radiation component into account for ho (outside tubes). Check pressure drop over designed heat exchanger in tube nest (process side).
(3) Make a sketch of the converter tower (see figure 10), taking into account that the catalyst beds have to be inspected annually and that sometimes the catalyst has to be removed for sieving or partial or complete replacement. Input data Appendix 3: Safety data and handling of Monsanto Vanadium Pentoxide catalyst.
Note that pressure drop measurements over each bed are required and that temperatures have to be monitored before and after each bed (see table 4).

12.3.4 SO_3/air - cooling air heat exchangers

Two identical vertical shell-and-tube heat exchangers in series are required for the cooling of the SO_3/air mixture leaving the converter tower. SO_3/air through tubes (say 15 mm internal diameter), cooling air through shell, cooling air inlet temperature to each heat exchanger 25°C. The inlet temperature of SO_3/air in the first heat exchanger is 435°C, the outlet temperature of the SO_3/air leaving the second heat exchanger should be maximum 50°C. Cooling air outlet temperatures for both heat exchangers are not fixed.

Problems to be solved

----

(1) Design the two identical shell-and-tube heat exchangers. Use relevant literature for design of shell-and-tube heat exchangers; take radiation contribution into account in first heat exchanger. Use mass balance figures from 12.2. and use mean Cp values (kJ/kg°C) for SO_3/air and cooling air in the temperature ranges in both heat exchangers.

(2) A small amount of 20% oleum will condense in the second heat exchanger (see 12.2.). Make a sketch of the second heat exchanger with oleum collecting system.

12.3.5 Cascade of stirred tank reactors for sulphonation (see 5.5.2.)

In table 13 an example is given for a plant with roughly 3 times the capacity of this example (12.2.) The conversion distribution should remain identical.

Problems to be solved

----

(1) Calculate the heat flow per reactor using a LAB flow of 710 kg/h or 3 kmol LAB/h (MW 237).

(2) Calculate the air flow leaving each reactor. Calculate the diameter of the first reactor assuming a superficial velocity of the depleted air in the top of the reactor of 0.45 m/s. All reactors will have this diameter.

(3) The cooling surface areas will become 6 m^2, 4.5 m^2, 3.8 m^2, 1.1 m^2. Assume that each reactor has 3 parallel cooling coils, each with an internal tube diameter of, say, 25 mm and a maximum cooling water velocity of 2 m/s. The inlet cooling water to each coil is 25°C. The maximum sulphonation temperature in each reactor is 65°C. The overall heat transfer coefficient for each reactor is given in table 13. Calculate for each reactor the actual sulphonation temperature, assuming a total heat of reaction of -170 kJ/gmol, the given conversion distribution over the reactors (table 13) and the above cooling water arrangement.

(4) Make a sketch of the cascade of the 4 reactors with gravity overflow from one to the next, leaving room and height for the ageing tank and stabilising reactor (see 5.5.10.) and a final stirred-tank neutraliser reactor.

12.3.6 Paste pumping

After neutralisation, the 55% AM paste is stored in a finishing tank where final quality checks are carried out before pumping the paste to the tank farm (100 m^3 paste tanks). The distance between the sulphonation plant and the tank farm is 120 m. Further data: 55% paste rheology data at 60°C (see table 16) K = 50 Ns/m^2, n = 0.2, paste density about 1000 kg/m^3.

Problems to be solved

----

(1) Which pipe diameter would you select for pumping the paste from the sulphonation plant finishing tank to the 100 m^3 paste storage tank, assuming a maximum pressure drop of 6 bar? The difference in height between the finishing tank outlet and the top of the paste storage tank in the tankfarm is 8 m. For pseudoplastic liquors like paste the pressure drop due to friction losses is given by the following equation:

$$\Delta P = 2KL/R \times \{[(3n+1) \times \phi v]/[\pi n R^3]\}^n$$

in which
- ΔP = pressure drop (N/m^2)
- K = consistency, that is the apparent viscosity at a shear rate of 1 s^{-1} $N.s^n/m^2$
- L = pipe length (m)
- R = radius of the tube (m)
- n = power law index, for pastes in the range of 0.2 - 0.3
- ϕ_v = volumetric flow [m^3/s]

The pipe should have steam tracing and should be lagged.

(2) What kind of pump would you select to pump about 2 tons of paste per hour to the tank farm?

(3) Prepare a simple P&I for the paste transfer arrangement (finishing tank with low and high level alarms, sample points, drain system, facilities to blow the pipe system with LP steam, high level alarm in the store tank).

12.3.7 Scrubbing of residual SO_2 from exhaust gas (see figure 44 and table 19)

From the mass balance (12.2.) the SO_2 and air quantities entering the scrubber are known.

Problems to be solved

----

(1) Prepare a mass balance for the caustic scrubber taking into account a total solids concentration (w/w%) of 10% and a pH of the exit liquor of 9-10.

(2) Calculate the height of the 1" Raschig Ring bed required for the following conditions:
- height of a transfer unit 0.3 m;
- SO_2 concentration in exhaust gas leaving the stack 5 mg/m^3 at a SO_2 conversion of 80%. This latter condition takes into consideration the plant start-up running conditions.

(3) What construction material would you select for the scrubber, packing material and pipeline system?

(4) Prepare a simple P&ID for this caustic scrubber.

12.3.8 Miscellaneous further tasks

(1) Prepare simple P&ID's for the following plant sections:
  a. process air drying
  b. molten sulphur proportioning
  c. SO_3 production
  d. sulphonation, ageing, stabilisation and neutralisation
  e. exhaust gas treatment

(2) Prepare a lay-out for the sulphonation plant (see 10.2.).

(3) Make a list of required laboratory equipment for the plant laboratory.

(4) Imagine that your company has decided to install a sulphur-based SO_3/air sulphonation plant on the household detergent factory site where you work. It is your task to introduce this new technology to the following personnel:
- general production manager;
- safety officer;
- environmental manager;
- recruited team of supervisor and shift operators (one supervisor in day shift, two operators per shift);
- plant maintenance manager and plant maintenance engineers.

Prepare brief presentations for the different categories of technical staff.

(5) Prepare a list of safety equipment required for the total operation (see all hazard data sheets).

12.3.9 Mass balance exercise for 20% oleum sulphonation plant

Table 41 shows a typical 20% oleum plant mass balance. The assumptions for this mass balance are summarised in section 11.1. In figure 52, a phase diagram of sulphonic acid - sulphuric acid and water is depicted.

Problems to be solved

Check the figures in table 41.

Appendix 1

ABBREVIATIONS

AB	Alkylbenzene
ABS	Alkylbenzene Sulphonate
ABSA	Alkylbenzene Sulphonic Acid
AD	Active Detergent
AES	Alcohol Ether Sulphates
AM	Active Matter
AO	Alpha-olefin
AOS	Alpha-olefin Sulphonate
CSTR	Continuous Stirred Tank Reactor
EO	Ethylene Oxide
ESP	Electrostatic Precipitator
FAME	Fatty Acid Methyl Ester
FAME-SA	Fatty Acid Methyl Ester - Sulphonic Acid
FAMES	Fatty Acid Methyl Ester Sulphonate
FFR	Falling-Film Reactor
Free oil	Non Detergent Organic Matter
GRP	Glass Reinforced Plastic
LAB	Linear Alkylbenzene
LABS	Linear Alkylbenzene Sulphonate
LABSA	Linear Alkylbenzene Sulphonic Acid
LES	Lauryl Ether Sulphate
MMW	Mean Molecular Weight
MT-FFR	Multi Tube - Falling Film Reactor
MW	Molecular Weight
NSD	Non Soapy Detergents
PA	Primary Alcohol
PAS	Primary Alcohol Sulphate
PEG	Polyethylene Glycol
PPM	Part Per Million
PTFE	Polytetrafluorethylene
SA	Sulphonic Acid
SFAS	Sulphonated Fatty Acid Sodiumsalt
TAS	Tallow Alcohol Sulphate

APPENDIX 2 Conversion table for physical properties

Magnitude	Expressed in	Multiply by → / ← Divide by	In SI units
Length	inch (in)	0·0254	m
	foot (ft)	0·305	
	yard (yd)	0·914	
	mile	1609	
	Angstrom (Å)	10^{-10}	
Area	in^2	$6{\cdot}45 \times 10^{-4}$	m^2
	ft^2	0·0929	
	yd^2	0·836	
	acre	4047	
	$mile^2$	$2{\cdot}59 \times 10^6$	
Volume	in^3	$1{\cdot}64 \times 10^{-5}$	m^3
	ft^3	0·0283	
	yd^3	0·765	
	UK gallon	$4{\cdot}55 \times 10^{-3}$	
	US gallon	$3{\cdot}785 \times 10^{-3}$	
Time	minute (min)	60	s
	hour (h)	3600	
	day	$8{\cdot}64 \times 10^4$	
	year	$3{\cdot}16 \times 10^7$	
Mass	grain	$6{\cdot}48 \times 10^{-5}$	kg
	ounce (oz)	$2{\cdot}84 \times 10^{-2}$	
	pound (lb)	0·454	
	hundredweight (cwt)	50·8	
	ton	1016	
Force	poundal (pdl)	0·138	N
	pound force (lbf)	4·45	
	dyn	10^{-5}	
	kg force (kgf)	9·81	
Volumetric flow	ft^3/min	$4{\cdot}72 \times 10^{-4}$	m^3/s
	UK gal/min	$7{\cdot}58 \times 10^{-5}$	
	US gal/min	$6{\cdot}31 \times 10^{-5}$	
Mass flow	lb/min	$2{\cdot}10 \times 10^{-6}$	kg/s
	ton/h	0·282	
Density	lb/in^3	$2{\cdot}77 \times 10^4$	kg/m^3
	lb/ft^3	16·0	
Pressure	lbf/in^2	$6{\cdot}89 \times 10^3$	N/m^2
	lbf/ft^2	47·9	
	dyn/cm^2	0·1	
	kgf/cm^2 (= at)	$9{\cdot}81 \times 10^4$	
	atm (standard)	$1{\cdot}013 \times 10^5$	
	bar	10^5	
	in water	$2{\cdot}49 \times 10^2$	
	ft water	$2{\cdot}99 \times 10^3$	
	in Hg	$3{\cdot}39 \times 10^3$	
	mm Hg (torr)	$1{\cdot}33 \times 10^2$	

Magnitude	Expressed in	Multiply by → / ← Divide by	In SI units
Dynamic viscosity	lb/ft h	$4{\cdot}13 \times 10^{-4}$	Ns/m^2
	lb/ft s	1·49	
	Poise (P = g/cm s)	0·1	
	Centipoise (cP)	10^{-3}	
Kinematic viscosity	ft^2/h	$2{\cdot}58 \times 10^{-5}$	m^2/s
	Stokes (S = cm^2/s)	10^{-4}	
	Centistokes (cS)	10^{-6}	
Surface tension	dyn/cm (= erg/cm^2)	10^{-3}	N/m
Temperature difference	degree F (or R)	5/9	°C (°K)
Energy (work, heat)	ft lbl	0·0421	J = Ws = Nm
	ft lbf	1·36	
	BTU	$1{\cdot}06 \times 10^3$	
	CHU	1899	
	hph	$2{\cdot}68 \times 10^6$	
	erg	10^{-7}	
	kgf m	9·81	
	kcal	$4{\cdot}19 \times 10^3$	
	kWh	$3{\cdot}60 \times 10^6$	
Power (energy flow)	BTU/h	0·293	W
	CHU/h	0·528	
	ft lbf/s	1·36	
	hp (British)	746	
	hp (metric)	736	
	erg/s	10^{-7}	
	kcal/h	1·163	
	cal/s	4·19	
Heat flux	BTU/ft^2 h	3·15	W/m^2
	cal/cm^2 s	$4{\cdot}19 \times 10^4$	
	$kcal/m^2$ h	1·163	
Specific heat	BTU/lb °F	$4{\cdot}19 \times 10^3$	J/kg °C
	kcal/kg °C	$4{\cdot}19 \times 10^3$	
Latent heat	BTU/lb	$2{\cdot}33 \times 10^3$	J/kg
	kcal/kg	$4{\cdot}19 \times 10^3$	
Heat conductivity	BTU/ft h °F	1·73	W/m °C
	cal/cm s °C	$4{\cdot}19 \times 10^2$	
	kcal/m h °C	1·163	
Heat transfer coefficient	BTU/ft^2 h °F	5·68	
	cal/cm^2 s °C	$4{\cdot}19 \times 10^4$	
	$kcal/m^2$ h °C	1·163	

Appendix 3

Safety data and handling of Monsanto vanadium pentoxide catalyst

A	Catalyst safety data
B	Instructions for installation of Monsanto sulphuric acid catalyst
C	Instructions for vacuum screening of catalyst

A SAFETY DATA

I. General warnings & precautionary measures

Dust may be harmful if inhaled or swallowed. Irritating to skin and eyes.

- Avoid breathing dust;
- Avoid contact with eyes and skin;
- Use adequate ventilation;
- Wear approved respirator;
- Wash thoroughly after handling;
- Emptied container retains dust, therefore observe above handling requirements until destroyed.

II. Occupational control procedures

a.	Eye protection	Suitable to prevent eye contact
b.	Skin protection	Wear appropriate impervious gloves and protective clothing to prevent skin contact. Contaminated protective equipment to be cleaned before re-use.
c.	Respiratory protection	Approved respiratory equipment. This will need to be a full face mask and it is recommended that a positive pressure system is used to give a good level of protection as well as a reasonable level of worker comfort. This will also provide protection to face and eyes.

Severe exposure may result in pulmonary edema and pneumonia. Following exposure, individuals may experience persistent bronchitis resembling asthma and bouts of dyspnea. No chronic lesions have been reported.

III. Silica-quartz

Crystalline forms of silica are fibrogenically active. The symptoms (chronic, usually after 10 - 25 years exposure) include dyspnea, cough, chest pain, decreased vital capacity etc. The lung fibrosis is irreversible.

IV. Emergency and First Aid procedures

a.	Eyes	Flush with water, seek medical attention
b.	Skin	Wash with water, remove contaminated clothing. Wash clothing before re-use.

c. Inhaled Remove from exposure, seek medical attention.

V. Fire Protection and Storage

Product is not flammable, but may evolve sulphur trioxide gas if heated. The catalyst should be kept dry to avoid material degradation.

VI. Summary of Effects Data

Vanadium Pentoxide

Vanadium pentoxide is irritating to respiratory tract, eyes and skin. Inhalation of dust or fumes may produce coughing, sore throat, irritation of the nasal passage, nose bleeds, inflammation of the trachea and bronchi, increased sputum and chest pain.

Symptoms of eye irritation include profuse tearing and a burning sensation of the conjunctiva.

Skin irritation is of the eczematious type with sensation of burning or itching.

Exposed individuals may exhibit a greenish discolouration of the tongue.

Monsanto

MATERIAL SAFETY DATA SHEET

SULPHURIC ACID CATALYST
Types 516, 210, 11, LP-120, LP-110

CHEMICAL COMPOSITION	Inorganic salt mixture containing 5-8% of V_2O_5 on a crystalline silica support and 10% of absorbed sulphur trioxide.
PHYSICAL FORM	Yellow greenish pellets and rings and up to 0.1% dust on despatch.
HAZARD WARNING LABEL	X_n - Harmful by inhalation and if swallowed. Irritating to eyes, respiratory system and skin.
TOXICOLOGICAL INFORMATION	Harmful by inhalation and if swallowed. (Acute oral toxicity LD_{50} (rat) 1460 mg/kg.) Practically non-harmful by skin absorption. (Acute dermal toxicity LD_{50} (rabbit) > 5010 mg/kg.) Slightly irritating to eyes (rabbit). Prolonged inhalation of crystalline silica may lead to impairment of lung function and silicosis. Practically non irritating to skin.(rabbit) Threshold Limit Value (TLV). V_2O_5 (Vanadium pentoxide)) 0.05mg/m^3 respirable ACGIH SiO_2 (Crystalline silica) (25-40%) 0.6mg/m^3 total dust 0.2mg/m^3 respirable (Amorphous silica) (15-25%) total 10 mg/m^3 respirable 5 mg/m^3 Human Experience Occupational exposure to these catalysts has been reported to cause severe eye, skin and upper respiratory tract irritation. Excessive acute exposure to vanadium compounds may produce coughs, wheezing, difficulty in breathing, chest pains, bronchitis, influenza-like symptoms and possible pneumonia.
FIRE HAZARD	If exposed to fire from another source use extinguishing agent for that material. Does not burn but when very strongly heated decomposes to give off toxic fumes (SO_3 above 430°C). Firefighters : wear self-contained breathing apparatus and full protective clothing. Thoroughly clean equipment after use.

01/88
(Replaces 11/86)
JSH

EMERGENCY FIRST AID

In case of contact with skin wash immediately with plenty of water.
Eyes : flush immediately with plenty of water for at least 15 minutes. Seek medical advice if irritation persists.
Inhalation of dust: remove patient to fresh air. Treat symptomatically.
If irritation persists seek medical advice.
Remove contaminated clothing and wash before re-use.

PHYSICAL PROPERTIES

Pellets : 5.6 or 8.0 mm diameter
: 9.5 mm average length
(Types 516, 210 and 11)
Rings : 12.5 or 9.5 mm diameter
: 12.5 mm average length
(Types LP-120 and LP-110)
Bulk density: 0.6-0.7 (Types 516,210 and 11)
0.5-0.55 (Types LP-120 and LP-110)
Slightly soluble in water.
Contains up to 0.1% dust on despatch.

HANDLING PRECAUTIONS

Provide local exhaust ventilation at source of contamination.
Avoid contact with eyes. Wear chemical goggles.
Have eye bath immediately available at any location where eye contact can occur.
Avoid prolonged or repeated contact with skin.
Do not breathe dust. Wear appropriate dust respirator.
Wear suitable protective clothing and gloves.
Wash contaminated clothing before reuse.

STORAGE

Keep container tightly closed and dry to avoid loss of activity of product. This will be reflected by a change from gold to green.

SPILLAGE, LEAKAGE AND DISPOSAL

Shovel up into covered containers for disposal.
Consult Monsanto for specialist disposal advice.

B. INSTRUCTIONS FOR INSTALLATION OF MONSANTO SULFURIC ACID CATALYST

I. General

Although shipping containers for Monsanto Vanadium Sulphuric Acid Catalyst are tightly sealed, they should always be stored in a dry place as the catalyst may be damaged by moisture or water. The containers should never be opened until the converter is ready to be packed.

It may be noted here that the colour of the pellets, as received, is not important. The pellets are sufficiently hard to withstand all normal handling and screening for dust removal after shipment usually will not be required. However, if the catalyst has been subjected to unusual shipping conditions, a small quantity of dust may be formed. Since dust will tend to raise the resistance in the converter, the pellets should be carefully screened before installation of catalyst, if found necessary by inspection. This screening is best done by passing the pellets gently over a screen of about 8-mesh (linear) size as the containers are opened for converter packing. Workmen handling the catalyst should wear respirators as the dust may be irritating to the throat and lungs.

Screening of catalyst (if required) and packing of the converter should be completed expeditiously so as to limit the duration of exposure to moist air.

In preparation for installation of catalyst, the inside of the converter, the converter cover and all of the miscellaneous fittings used inside the converter should be brushed clean of scale and dirt with a wire brush and care should be taken that workmen clean their shoes before entering the converter. All interior fittings should have been assembled in the converter before beginning the installation of the catalyst, to be sure that, once begun, the installation can proceed without undue delay for the fitting of parts.

After the slotted grids have all been installed, they should be pushed toward the center of the converter and fitted together so that it is not possible for any of the 1/2" quartz pebbles to fall through. Then two or three strands of insulating rope may be used to caulk around the shell of the converter. No caulking should be used between the slotted grid sections as they must be placed to allow for some expansion.

Occasionally it may be found that the converter shell is not truly cylindrical throughout or that excessive openings occur between the grids. In such cases, it is suggested that an excessive opening between grids be closed by inserting a steel tee bar. Excessive openings around the shell may be closed by spot welding an angle or bar to the shell immediately under the grid and then filling the space with asbestos rope.

The solid partition plates must be carefully caulked between each plate and between the plates and the converter shell with several strands of insulating rope. It is extremely important that no gas be able to pass these solid partition plates. Gas leaks past these solid partition plates will cause poor temperature control and result in loss of conversion efficiency.

The catalyst supporting medium in each layer consists of 1/2 inch by 1 inch milled quartz pebbles which should not spall at 600°C (1112°F). A 2 inch deep layer is put on each set of slotted grids and, after being levelled carefully, is followed by the specified volume of catalyst pellets. It is suggested that a piece of 2 inch thick plank be used as a depth guide for the bottom layer of quartz. A similar layer of quartz pebbles is also placed on top of each layer of catalyst.

Catalyst should be handled gently and not thrown in recklessly or dumped from a great height. It is best handled in pails or in the original containers (after rescreening, if found necessary). After the specified amount of catalyst has been placed, the layer should be carefully levelled to insure uniform thickness in the catalyst bed. Failure to level properly will cause excessive flow through "thin spots" and may result in poor conversion. It has been found convenient to make three chalk marks around the columns and around the wall of the converter at a uniform distance above the grid. The bottom chalk mark represents the approximate top level of catalyst. The second chalk mark, two inches above the first, represents the anticipated level of the quartz pebbles over the catalyst. The third mark is placed two inches above the second mark and serves as a reference if the second mark is covered by quartz pebbles.

Any of the bottom layers (except as noted for wet gas purification plants) may be packed through access nozzles on the converter shell, without disturbing the others. Wide boards or panels of plywood should be used as a working platform while levelling the catalyst and installing the top layer of quartz. In no case should men be permitted to step or lie directly on the catalyst. Care must be taken at all times that dirt, pieces of wood, cloth, etc. do not get into the catalyst or the converter.

In cool, damp weather it is desirable that electric heat lamps or large electric light bulbs be used in the converter to aid in keeping it dry while installing catalyst. Some types of heat lamp will serve the dual purpose of supplying both heat and illumination. Before sealing on the manhole covers, inspect to determine that all equipment has been removed and that all parts or the converter are clean. Remove any dust that may have fallen through the grids on to the division plates while installing the catalyst.

II. Converters in sulphur burning plants

Sulphur-burning plants with waste heat boilers frequently have superheater tubes in the converter. In smaller plants, the tubes of one of the boiler sections may also be inside the converter. There is usually but a few inches' clearance between the baffle plate under the tubes and the catalyst. When installing catalyst in the layers containing these tubes, it is best to use the manholes on each side of the converter. This will facilitate the final levelling of the catalyst and the covering layer of quartz pebbles.

Any thermocouple protecting wells inside the converter must be located in their proper places while the catalyst is being installed. They should be located exactly at the dividing line between the catalyst and the quartz pebbles in each instance as shown on the drawings. A handful of quartz may be placed under the end of the thermocouple well to prevent the tip of the thermocouple from being buried in the catalyst.

C. INSTRUCTIONS FOR VACUUM SCREENING OF CATALYST

It may be necessary at some time to clean the catalyst in the converter because of increased pressure drop due to dust and scale from various sources being carried in the gas stream to the converter. In sulphur-burning plants, impurities in the sulphur will cause catalyst plugging.

1. Cool the converter by blowing dried air through it until the temperature is low enough for men to work. Normally, 50 -65°C is sufficiently low, as when the converter is opened up, it will cool rapidly to the point where it can be worked in comfortably. If minimum shutdown time is desired, two manhole covers on the converter may be removed and an exhaust fan installed in one while the work is done through the other.

2. Prior to shutting down, erect a temporary platform at the converter manhole from which catalyst will be removed. Provide safe access to the platform.

   The thermocouples and wells should be removed before removing the catalyst so that they will not be damaged. Be sure that the thermocouples are installed before the catalyst is replaced, being certain that they are replaced at the proper location. Thermocouples should always be located at the rock/catalyst interface.

   The screening operation should never be conducted while it is raining or when rainfall impends. Polyethylene sheeting and rope or wire should be on hand to cover the manholes during rainfall and when screening work is not taking place.

3. In general today, most catalyst is removed from the converter by vacuum. The catalyst is collected in a cyclone-like vessel and fed by feeder on to a two-layer mechanical screen which separates the quartz from the catalyst and the catalyst from the dust in one operation. In some instances, it is desirable to separate pellet catalyst from ring catalyst, so that the ring catalyst can later be reinstalled on top to maximize dust penetration which minimizes pressure drop build up.

   The quartz and catalyst are collected in plastic bags or some other type of watertight containers stored until ready to return to the converter. Catalyst should be stored indoors or protected outdoors in unfavourable weather conditions.

4. Prior to removing catalyst and rock, draw a line around the converter shell at the rock level. When the converter bed is empty, draw another line 2" below the original one. These will serve as guides for reloading to the proper height.

5. Before the catalyst layer is repacked, thoroughly clean the division plate under the grids. This may be done from manholes if provided, or by removing an occasional grid to reach this area. The use of a vacuum cleaner is desirable, but brushing with a soft brush gives a satisfactory result. Any dust or dirt left on the division plate may blow into the next layer of catalyst. Observe the condition of the grid slots and clean out if they are plugged with bits of quartz and catalyst. Normally, these grid slots will not require cleaning, but cases have been known where, after several cleanings and rough handling of the quartz, small pieces of quartz have

become lodged in these slots causing added pressure drop. It is recommended that an inspection be made of caulking between the grids to make certain there are no openings large enough to allow quartz and catalyst to fall through the grids. Some division plates are covered with insulating brick. This insulation brick should be inspected and repaired if necessary.

6. Normally, catalyst cleaning is contracted out and the contractor will supply the vacuum unit with hoses, screens, etc. and a crew to operate the unit. The plant owner will then supply a crew with a supervisor to handle the catalyst removal and placing in containers.

7. On reinstalling catalyst, be sure a good 2" layer of quartz is installed on the grids and use wide boards or panels of plywood as a working platform inside the converter. In no case should men be permitted to walk directly on the quartz or catalyst, as this will cause holes in the quartz supporting the catalyst and will also break the catalyst pellets.

8. The manholes should be closed and the converter sealed off from all access by water or wet air until the plant is restarted.

Revised 3/15/83 W.E. Poncez
Monsanto Enviro Chem., P.O. Box 14547, St. Louis - Missouri 63178

Appendix 4 DETAILED ANALYTICAL METHOD REFERENCES (source Ballestra)

Detailed analytical method references are given below. The data were supplied by Ballestra SpA, Milano.

Legend:

ASTM	American Society for Testing Materials
AOCS	American Oil Chemists' Society
BS	British Standards (Methods of sampling and testing detergents)
IP	IP standards for petroleum and its products (methods for analysis and testing)
SMWW	Standard methods for the examination of Water and Wastewater
DEWAS	Deutsche Einheitsverfahren zur Wasseruntersuchung
BAL	Ballestra numbering
NST	Norm Sostanze Tensioattive (Norms for detergents)
NGD	Norme Grassi e Derivati (Norms for oils and fats)
NOH	Norme Olii Minerali e Derivati (Norms for mineral oils and derivatives)
L	Longman: The analysis of detergent and detergent products
R	Rosen & Goldsmith: Systematic analysis of surface-active agents, 2nd edition (for both: J. Wiley - interscience publication)
D	Milwidsky & Gabriel: Detergent analysis (editor: G. Godwin)

1. Alkylbenzene

BAL 1.1.1.	**Water**	NST BC IV - 1 NOM 117-80 ASTM D 1533-79 AOCS Dd 2a-59 AOCS Dd 2b-59 L 435	AOCS Tb 2-64 AOCS Tb 2a-64 R 329
BAL 1.1.2.	**Mercaptans**	NST Bc III-1	
BAL 1.1.3.	**Bromine number**	NST Bc IV-2 IP 129-64 IP 130-67 ASTM D1159-66	D 193
BAL 1.1.4.	**Distillation test**	NST Bc II-3 NOM 15-71 IP 123/68 IP 191/65	ASTM D 86-67 ASTM D 216-54

2. Fatty alcohols

BAL 1.2.1.	**Water**	[BAL 1.1.1.]	
BAL 1.1.2.	**Acid value**	NST Ba IV-2	
BAL 1.2.3.	**Saponification number**	NST Na IV-3 NGD C33-76 AOCS Cd 3-25	
BAL 1.2.4.	**Iodine number**	NST Ba IV-4 NGD C32-76 AOCS Cd 1-25 D 192	
BAL 1.2.5.	**Hydroxyl number**	NST Ba IV-5 NGD C34-76	

3. Ethyloxylated alcohols

BAL 1.3.1.	**Water**	[see BAL 1.1.1.]	
BAL 1.3.2.	**Polyglycols**	NST Eb IV-3 L 312	
BAL 1.3.3.	**Ethylene Oxide**	NST Eb IV-2 L 288	

4. Sulphur

BAL 1.5.1.	Sampling
BAL 1.5.2.	Moisture
BAL 1.5.3.	Bitumen
BAL 1.5.4.	Ash
BAL 1.5.5.	Sulphur
BAL 1.5.6.	Arsenic
BAL 1.5.7.	Chlorine
BAL 1.5.8.	Sulphuric acid
BAL 1.5.9.	Sulphurous and hyposulphurous acid
BAL 1.5.10.	Selenium

5. Water

BAL 1.6.1.	**Hardeners**	SMWW 309B DEWAS E4	
BAL 1.6.2.	**Alkalinity**		
BAL 1.6.3.	**Sulphates and chlorides**		

6. Alkylbenzene sulphonates

BAL 2.1.1.	**Colour**		
BAL 2.1.2.1.	**Water**	NST Cc IV-1 NGD C4-76 ASTH D460-60 BS 1715-51 API 2560 L 438 L 546 ASTM 095-62	AOCS Dc 2-59 AOCS Da 26-42 AOCS Dd 2a-59 AOCS F 1a-44 NOM 7-70 R 329

BAL 2.1.2.2.	**Unsulphonated matter**		
		NST Cc IV-3	
		AOCS Dd 4-60	
		[see BAL 2.2.1.2.]	
BAL 2.1.2.3.	**Colour**		
BAL 2.1.2.4.	**A.M. (p-Toluidine)**		
		L 249	R 392 (396)
	Hyamine	NST Gb IV-14	
		L 234	
		R 427	
BAL 2.1.2.5.	**Sodium chloride**		
		NST Cc IV-5	
		L 445	
		NST Cb IV-5	
BAL 2.1.2.6.	**Sulphates**		
		L 442	

7. Alkyl sulphate

BAL 2.2.1.1.	**Water**	NST Cb IV-1 [see BAL 2.1.2.1.]
BAL 2.2.1.2.	**Unsulphonated matter**	
		NST Cb IV-3 AOCS Dc 8-59 [see BAL 2.1.2.2.]
BAL 2.2.1.3.	**Colour**	[see BAL 2.1.2.3.]
BAL 2.2.1.4.	**A.M.**	[see BAL 2.1.2.4.]
BAL 2.2.1.5.	**NaCl**	NST Cb Iv-5 [see BAL 2.1.2.5.]
BAL 2.2.1.6.	**Na_2SO_4**	[see BAL 2.1.2.6.]

8. Ethoxy-alkyl sulphates (for dioxane analysis see 10.)

BAL 2.3.1.1.	**Water**	[see BAL 2.1.2.1.]
BAL 2.3.1.2.	**Unsulphonated matter**	
		H ls

BAL 2.3.1.3.	**Colour**	[see BAL 2.1.2.3.]
BAL 2.3.1.4.	**A.M.**	[see BAL 2.1.2.4.]
BAL 2.3.1.5.	**NaCl**	[see BAL 2.1.2.5.]
BAL 2.3.1.6.	**Na_2SO_4**	[see BAL 2.1.2.6.]

9. Analysis for FAME/FAME-SO_3H/FAME-SO_3Na

Phase	Analysis	Rif. methods
FAMES	SAP NUMBER	AOCS Da - 16-48
	H_2O CONTENT	ISO - 760
	J_2 VALUE	AOCS - Da 15-40
	UNSAPONIFIABLE MATTER	AOCS - Da 11-42
	COLOUR	(NGD C20-1976 (AOCS Cc - 136-45
FAME-SO_3/H	ACID VALUE	AOCS Da - 14-48
	H_2SO_4	Ballestra Standard Method(*)
	COLOUR	Ballestra Standard Method(*)
FAME-SO_3Na	H_2O	AOCS - Da 2B-42
	UNSULPHONATED MATTER	AOCS - Dc 8-59
	SAPONIFIED MATTER	AOCS - Db 6-48
	COLOUR	Ballestra Standard Method(*)
	A.M. (Hyamine)	ASTM - D - 3049-75
	NaCl (Volhardt)	AOCS - Da 9-48
	Na_2SO_4	Ballestra Standard Method(*)
	NON ESTER ACTIVE MATTER	Ballestra Standard Method(*)

10. Determination of 1,4-dioxane in ether sulphates by headspace gas chromatography

Principle
The content of 1,4-dioxane in ether sulphates is determined by headspace gas chromatography according to the standard additions method.

The method is suitable for all ether sulphates and gives reliable results independent of chainlength distribution and water content.

The time required for one determination is 1 1/2 hour.

Reagents
N,N-dimethylformamide (DMF), highest purity
1,4-dioxane, analytical reagent grade

Apparatus
Conventional laboratory equipment
Screw-cap jars 50, 250, 500 ml
Crimp-top glass vials 8 ml
Gas chromatographic system (e.g. Sigma 2, PE)
Headspace sampler (e.g. HS-6, PE)
Hand crimper

Measuring component

Detector	FID
Column	SINOX SH steel, length 3.6 m, ID 2.6 mm
Column packing	20% SP 2100 and 0.1% Carbowax 1500 on Supelcoport, 100/120 mesh
Carrier gas	Nitrogen, 40 ml/mln
Injector	Splittless
Injector temperature	200°C
Oven temperature	110°C, isothermal for 12 minutes
Detector temperature	250°C
Sample temperature	70°C
Equilibration time	30 minutes
Pressurising period	8 minutes

Calibration and standardisation
1,4-dioxane standard solutions:

I. Weigh 1.20 g 1,4-dioxane and 398.80 g DMF into a 500 ml screw-cap jar and homogenise carefully.

e.g. 5.0 g solution I mixed with 15.0 g sample = 1000 g/g
10.0 g solution I mixed with 15.0 g sample = 2000 g/g

II. Dilute solution I with DMF to get standard solutions which are within a range of 10 to 1000 g/g for a sample weight of 15.0 g.

Generate a calibration curve with these standards to calculate the appropriate content of 1,4-dioxane content in the sample.

Warning - due to matrix effects, this calibration curve cannot be used to calculate the real 1,4-dioxane content in the sample.

Sampling and sample preparation
Weigh 15.0 g of the sample into a 50 ml screw-cap jar, add 5.0 g DMF and homogenise carefully.

If the mixture is not homogeneous, weigh in another 15.0 g portion and add 10.0 g DMF.

Weigh 2.0 g of this mixture into a crimp-top glass vial and close it by a teflon septum.

In the same manner a second sample is prepared, but instead of pure DMF a suitable standard solution is used to double the peak area obtained for the previous sample.

Procedure
Chromatograph the sample vial and the stacked sample vial (para 6) according to the measuring conditions (para 4).

Calculation
The 1,4-dioxane content can be calculated by the following equation:

W (g/g) = (A x B)/(C-B)

where:
W = 1,4-dioxane content in g/g
A = stacked 1,4-dioxane in g/g
B = peak area of 1,4-dioxane in the sample
C = peak area of 1,4-dioxane in the stacked sample

Appendix 5

Sulphonation Plant Troubleshooting (source Chemithon)

Symptom:	Causes:
Low process air flow	- Damaged main compressor - Plugged dryer beds - Relief valve pop-off - Plugged oleum filter - Plugged filters - Plugged lines - Air dryer valve leaks
Low sulphur flow rate	- Damaged/dirty sulphur pump - Sulphur quality - Plugged Sulphur filters - Improper sulphur temperature - Instrument problems
Fluctuating burner outlet temperature	- Sulphur flow problems - Air flow problems - Erratic sulphur burning
High process air dew point	- Regeneration temperature/time - Air chiller failure - Air dryer control problem - Instrumentation failure - Damaged desiccant
Gas plant temperatures not correct	- Valve failure - Cooling air/water flow leaks/plugs in lines - Catalyst failure
Sulphonic acid product quality	- Mole ratio - Temperatures - Dirty reactor - Wet air/oleum carryover - Plugged filters - Gas strength - Poor organic/gas distribution - Reactor loading - Recycle pump failure - Feedstock quality

Poor neutralized product quality	- Acid quality - Flow rates - Mixing pump - Temperatures - Seals - Raw materials
Sulphuric acid absorber poor scrubbing	- Acid strength - Recycle pump failure - Mist eliminator pad - Packing failure
Poor SO_3 scrubbing in sulphuric acid absorber	- Temperature
ESP not removing particulate	- T/R set problems - Dirty
SO_2 scrubber not removing SO_2	- Caustic strength - Organic carryover - Precipitation - Packing failure

Subject Index